PAST, PRESENT
AND
FUTURE

PAST, PRESENT, AND FUTURE

ISAAC ASIMOV

PROMETHEUS BOOKS
Buffalo, New York

91 90 89 88 87   5 4 3 2 1

Library of Congress Cataloging-in-Publication Data

Asimov, Isaac, 1920-
    Past, present, and future.

    I. Title.
AC8.A72   1987   081   87-2243
ISBN 0-87975-393-5

To the memory of

John Ciardi (1916-1986)

TABLE OF CONTENTS

Part III: Future

PAST, PRESENT
AND
FUTURE

INTRODUCTION

Three years ago, I put together a collection of essays under the title of *The Roving Mind* for Prometheus Books. It was an utterly miscellaneous collection ranging, as I said in my introduction to the book, "from the polemical to the persuasive, from the speculative to the realistic."

That sort of thing is not the stuff of best-sellers, but the book did rather better than I thought it would. What is much more to the point, it did rather better than the publishers thought it would, and you know what that means.

Right. The suggestion was made by me (and quickly agreed to by Mr. Victor Gulotta of Prometheus) that I do another collection, and here it is. You find before you sixty-six articles, most of which were extorted from me by some kindly editor.

This collection is almost precisely the size of the first and, if anything, it is even more miscellaneous. It contains explanations of scientific points, speculations on the future, descriptions of various enthusiasms of mine, personal adventures and misadventures, and so on.

Again there are a few places where the essays overlap each other, or, in rare cases, contradict each other. Occasionally, there may be a bit of outdating for, even though almost all are quite recent pieces, things move quickly.

One warning, by the way. I plan to add an occasional foreword, or afterword, to point out anything I think of interest.

PART I
PAST

1

UNITY

I wrote the following essay in March 1985 in response to a request from the office of United States Senator Spark Matsunaga of Hawaii. He was planning a book, The Mars Project, in which he advocated a combined attempt by the United States and the Soviet Union to send a manned expedition to Mars.

In addition to the scientific benefits, there were the social and political benefits of such cooperation.

I heartily approved (and still do) the Senator's initiative. Citing examples in history, I gladly wrote an essay on the tendency for competing political units to combine and cooperate.

I was terribly disappointed, when the book was published, to find that the Senator had merely extracted a few paragraphs from the essay and published those.

However, I am never at a loss in such cases, for periodically I publish collections of my essays, and I knew that in the very next one I would publish, this essay, "Unity," would be included in full. And here it is.

As a matter of fact, while I have your attention, I might say that a number of essays in this collection were, in their original appearance in one outlet or another, edited to some trifling degree by the office staff of that outlet. That is their right; I have no objection. In this collection, however, the essays are pretty much as I wrote them. I have sometimes adopted the editors' suggestions, but not often. (I can't help it; I like what I write.)

So now let's go on to "Unity."

For almost all of history, human beings have been hunters of animal food and gatherers of plant food. In a hunting and gathering economy, the natural division is the extended family, or tribe. The tribe cannot be more than a relatively small group, for it must extend its search for a day's supply of food over the limited area that can be searched through in one day, on foot and without technological help. The group cannot be larger than the number that can be fed in a cropping, so to speak, of that limited area.

Population was and remained low, therefore, as small bands roamed the land. Ten thousand years ago, the total population of the Earth may have been eight million—about as many as in New York City today.

But then, through some gradual process of thought and serendipity, human beings learned how to prepare food for future use. Instead of hunting animals and killing them on the spot, some were kept alive and taken care of. Human beings became herdsmen, encouraging the animals to breed and multiply, and killing a few now and then. In this way, there was not only a secure supply of meat, but milk, eggs, wool, and muscular labor were attained as well.

Again, instead of just gathering what plant food could be found, human beings became farmers, learning how to sow plants and care for them so that they could be harvested at the proper time. Plants could be sown in much greater concentration than they could be found in nature, and population began to increase in the first "population explosion" of history.

Herding, however convenient, did not fundamentally alter the human way of life. Human beings no longer wandered in search of food by themselves; they wandered with their herds in search of pasture. Herdsmen, like hunters and gatherers, were nomadic.

It was agriculture that introduced the great change.

Farms, unlike human beings and animals, were not mobile. The need to take care of those farms nailed the farmers to the ground. The more they grew dependent upon the harvest to maintain their swollen numbers (too great for most to survive if they had to return to hunting and gathering), the more helplessly immobile they became. They could not run away from wild animals, nor could they from nomadic raiders who wished to help themselves to the copious foodstores that they had not worked for (and farming is back-breaking work). When the farmers *did* run away, they lost their food supply and would experience famine.

It followed that farmers had to fight off their enemies. They had no choice. They had to band together and build their houses in a huddle, for in unity there was strength. Often they had the good sense to build their houses on an elevation where there was a natural water supply, and to construct a wall about it. Thus were built the first cities.

Undoubtedly, there seemed something unnatural about this to the first city-builders. The natural unit seemed to be the family or tribe, and each tribe must have felt a certain antipathy toward banding together with other tribes (with "foreigners"). Suspicion, or even hatred, had to be overcome, and *was* overcome, for only in union was there safety. The need for help in case of emergency, the need of a refuge in case of threatened defeat, made the city essential, and a wider "patriotism" than that confined to relatives was clearly required.

Furthermore, when farmers were made reasonably secure thanks to their cities, they found that more food could be obtained than was required for their own needs. Some of the city dwellers, therefore, could do work of other types and exchange their products for some of the excess food produced by the farmer. The cities became the homes of artisans, merchants, administrators, priests, and so on. Human existence transcended the continual search for food, warmth, and shelter, and human beings began to specialize and to take on the kind of variety that we particularly value today. In short, civilization became possible, and the very word "civilization" is from the Latin for "city-dweller."

Each city developed into a political unit, with some system of government. The necessity of being prepared for battle against nomads led to the development of soldiers and weapons which, during peaceful periods, could be used to police and control the city population itself. Thus, there developed the "city-state."

As population grew, each city-state tried to extend the food-growing area under its control. Inevitably, neighboring city-states would collide with each other, and some firm boundary between them would have to be established. The difficulty was that, whatever the boundary, it was sure to seem unfair to some on each side; and even if there was general agreement, each side was sure to accuse the other of violating that boundary. Any vigorous king would see the advantage, moreover, of seizing his neighbor's property.

Thus, with the invention of the city-state came the invention of war. This is not to say there wasn't fighting in the old hunting and gathering days. When two tribal units encountered each other, there might well have been skirmishing, but the weaker side could always run away, so that the results weren't bloody.

City-states, however, were in fixed positions. They could not run away but had to stand and fight. Moreover, the fact that city-dwellers specialized and had time to think of things other than their next meal meant that weapons could be developed and improved. This was especially true once people learned to smelt ores for metals, since metals were much more efficient for the manufacture of tools and weapons than rock was. Wars grew progressively bloodier.

With the development of the city-state, then, human beings found themselves faced with an alternative: unity or war. To avoid war, potential antagonists had to combine in cooperation toward some common aim. The danger was that some group within the combination might try to dominate the others and might even succeed at it. Furthermore, the difficulty of administering a large unit might lead to a despotism, because the easiest way of coming to a decision on an issue when a group is too large to discuss it en masse is to let one man make the decision and then forbid further discussion or question.

To avoid despotism, people must take up the "small is beautiful" attitude and opt for political freedom. The result was, inevitably, war between the small, free units.

Throughout history, the pendulum has swung first this way, then that. If freedom seems to have become more precious with the ages, war has certainly grown more devastating. The stakes have grown enormously from generation to generation, and the decision has never yet swung clearly to one side or the other.

There were periods when the overwhelming need for cooperation swung the pendulum to the side of unity.

Early farming communities depended upon rain to supply the water needed for crop growth, but rain is notoriously fickle. The most successful communities were those that made use of land along the river banks. These developed dikes to confine the river and prevent ruin through floods. They also constructed a system of ditches that brought a controlled supply of water directly to the farms. The river, furthermore, served as an easy avenue for commerce, transportation, and communication.

It is for that reason that the earliest advanced civilizations developed along such rivers as the Tigris-Euphrates, the Nile, the Indus, and the Yangtze. Of these civilizations the Tigris-Euphrates was the first, and it was there that writing was invented—and the transition from "prehistory" to "history" took place—about 3200 B.C. By 2500 B.C., the Tigris-Euphrates was home to thirteen flourishing city-states.

However, to dike a river and maintain a system of irrigation ditches requires cooperation. If one city-state allows its own system to deteriorate, the flood that might follow would affect all the others disastrously. Even in war, a city-state would, for its own protection, have to protect the irrigation system of the neighbor it was trying to defeat.

Clearly, the prosperity of all required a true cooperative unity for the purpose of taming and controlling the river. Such unity, however, was not easy to attain in view of the suspicions and hatreds that separated the city-states. Each one could point to injustices perpetrated by others in the past and portray itself as an inoffensive victim of a wicked neighbor.

Nevertheless, between 2350 and 2300 B.C., a ruler named Sargon established the first empire. By fighting victorious wars he came to rule over the entire Tigris-Euphrates valley, and he became the first empire-builder.

It is probably a mistake to think that an empire of this sort is the result of nothing but the military aptitude of the victorious general. It is likely that in all cities there are those who understand the value of cooperation. With a single government controlling the entire valley, the irrigation system would surely be run with greater efficiency and that would mean prosperity. War between the cities would cease and that would mean security.

Yet such an empire cannot last forever. With the passing of time, prosperity and security are taken for granted and come to be undervalued. Instead, for each city the memory of independence becomes idealized so that there is longing for its return, a longing sufficient to bring about a breakup through repeated rebellions. And onslaughts of nomads from without break up the empire all the faster. Within a century, Sargon's empire was shattered.

What followed then in the Tigris-Euphrates was a repeating pattern of empire-building followed by empire-breakdown followed by empire-building again.

In Egypt there was little danger from external attack during the first two thousand years of its civilization. The desert protected the Nile River along which the civilization grew. Even so, after a strong central government was established, it gradually broke down into small independent regions until another capable leader formed a strong central government again.

It is easy to see that in this alternation of empire and small fragmentary communities, it is the empire that usually represents a time of prosperity and security. To be sure, the existence of the small-fragment stage may encourage cultural variety and, through competition, increase the advance of technology and the arts, but that stage invariably brings with it endless internecine warfare and before long the exhaustion of the land and the people.

The best example of the glory of the small-fragment stage is Greece in the fifth century B.C. It fairly exploded with art, science, and philosophy and defeated the Persian Empire—and then it committed suicide when it was unable to stop the useless unending wars that no one could win. In fact, the disadvantages of the small-fragment stage are so plain that it is frequently referred to as a "dark age."

Empires tend to expand after they are formed, for there is always the desire to push back the threatening nomads at the border or to weaken a dangerous neighboring empire. Such expansion always comes to an end, for eventually the technological level of a society makes it impractical to attempt to control an area that has grown over-large. It becomes impossible for the central authority to know in time the disasters and rebellions that take place at the periphery, and to be able to take corrective measures quickly enough.

Of course, as technology advances, the territory that can be controlled becomes larger. For that reason, as one empire succeeds another, the tendency is for the size to increase. The Persian Empire absorbed all the civilized area in western Asia and northeastern Africa in the sixth century B.C. and governed over an area comparable to that of the United States. This was made possible by a system of horse relays like the American Pony Express that knit the large area together. It was barely enough, though, and it fell apart after a mere two centuries under the blows of Alexander the Great. Nor did Alexander's empire survive his death.

The Roman Empire was as large as the Persian Empire and was far more successful. It was built up over the course of eight centuries and it declined and fell over the course of thirteen more centuries. At its fullest extent in the second century A.D., it was held together by a road system incomparable for the pre-industrial era. Many people are puzzled by the fact that the empire fell apart at all when it was so successful in so many ways. My own feeling is that it fell because it never developed a rational system of taxation, and that financial failure was the cause of its ruin.

The largest of all empires of the pre-industrial era was the Mongol Empire of the late thirteenth century, which included what is now the Soviet Union, China, and much of the Middle East, and that lasted for only fifty years before it disintegrated into rather large fragments.

When the western provinces of the Roman Empire broke up there followed what is to the western world the best-known of the Dark Ages. It took a thousand years for the fragments to pull together at least partially. France, Spain, and England formed. Germany and Italy remained as city-states and mini-nations and were victimized by the larger nation-states until they formed nations themselves in the nineteenth century. Europe had its golden age in the eighteenth and nineteenth centuries when it played a repeat of the role of Greece. Its nation-states were larger than the Greek city-states, but their rivalries and competitions led to the same flourishing of art and science, the same advance of technology that made it possible for them to subjugate large regions of other continents as Greece subdued the Persian Empire. But the nation-states of

Europe, like the city-states of Greece (and the medieval city-states of Italy), never ended the wars among them, and committed suicide as a result, and did it just as quickly as the Greek cities did.

Just as Greece finally lay quiescent under the large Macedonian states, and then under Rome; and just as Italy lay quiescent under France and Spain; Europe now lies quiescent under the Soviet shadow in the east and the American shadow in the west. Europe may still be prosperous, but its destiny is entirely in the hands of the superpowers. It is helpless when it tries to exert any meaningful effect of its own.

The industrialization of the world makes it possible for large areas to be held together with extraordinary efficiency. If we measure by the time it takes people to travel from one end of a realm to the other, nations like the Soviet Union, the United States, China, Canada, and Brazil are no larger than ancient city-states. Indeed, one can send a message from Washington to Moscow now much faster than one could send one from Athens to Sparta in the time of Pericles.

In essence, there is nothing in theory that can keep the whole world from being governed by some central authority. A true United Nations could keep tab on every corner of the world much more rapidly and effectively than an Egyptian Pharaoh could control five hundred miles of the Nile River.

From a practical standpoint, however, it would seem that the divisive influences in the world preclude such a possibility. Trifling differences in religion suffice to fuel irrational wars between Iraq and Iran, or between Protestants and Catholics in Northern Ireland. Israel is an indigestible problem to the Arab nations of the Middle East, and South Africa to the Black nations of Africa. Most of all, the competing economic and social outlooks of the Soviet Union and the United States seem to make it absolutely impossible for true international cooperation to come to pass.

And so, in the light of the present technological capacity of humanity, the planet is in a small-fragment stage. The Soviet Union makes up but one-sixth the land area of the world; China but one-fifth of the population; the United States but one-sixteenth the land area and one-twentieth of the population.

Until the twentieth century, such a situation would have been allowed to work itself out. Nothing could possibly have kept the world-fragments from fighting each other endlessly. World empires would have been established—till they fell apart.

That, however, cannot be counted upon now. Advancing technology has made war so frightful that it seems quite obvious that any attempt to select a winner in this manner would result in the destruction of civilization and, very possibly, most of life.

The same frightfulness has held serious war between the two superpowers in abeyance for forty years, but this only means that humanity and the world has lived under a Damoclean sword for all that time. Unless there is firm agreement to cooperate and live in peace the chance of war remains. And with time, through exasperation, through miscalculation, through accident, it will surely come and we will all be dead.

So there must be world unification—and yet it would seem that there can't be world unification. We can't force world unification by war, because we chiefly need world unification to *prevent* war.

Is there any way of bringing about world unification by some means other than war? Is there any example in world history of a unification among hostile fragments brought about without war?

Yes, there is!

When the United States completed its War of Independence and made peace with Great Britain in 1783, the new nation was anything but "united." It consisted of thirteen states, each of which jealously guarded its rights. There were differences among them in religion, in ethnic make-up, in governmental systems, and in social outlook. There was a national Congress, a national citizenship, a national post office, and so on, but there was no national taxation. Congress, without power, subsisted on what were virtually free-will offerings from reluctant states. The Congress was indeed an early analog to the present-day United Nations and no more powerful.

Had this situation continued, the United States would surely have fallen apart. Each state would have seized as much of the western territories as it could; there would have been economic and even military war among them (indeed, both were threatened on a number of occasions in those years). There would have been shifting systems of alliances and endless, inconclusive fighting. The American states would have committed suicide, and in the end the various states would have eventually become satellites of one European nation or another.

This is not just hindsight on my part. There were people who saw this clearly at the time, and who wanted a strong central government to prevent this, and to deal with issues that affected all the states alike (interstate commerce, the Indian threat, British refusal to evacuate some of its bases, and so on).

There were others, however, who saw in a strong central government the possibility of despotism and the loss of liberty—the same opposing fears of loss of security on one hand and loss of liberty on the other that have harrowed the world for over four thousand years now.

Yet a viable compromise between the two was worked out.

It began with the problem of the Potomac River and Chesapeake Bay, waters which were shared by Maryland and Virginia. But just how were the waters to be shared? James Madison of Virginia, later the fourth president of the United States, suggested Maryland and Virginia meet and negotiate some sort of treaty on the matter. Maryland suggested that other states join in. The meeting was held in Annapolis in 1786, but it fizzled. Delegates from only five states were present, and they sat for only four days. Alexander Hamilton of New York was there, however. An authentic genius, he saw very clearly the need for a strong central government, and he persuaded the delegates present not simply to disband but to call for another meeting. He volunteered to write the resolution himself.

He did, and he called for a convention to be attended by *all* the states to deal with *all* the problems that beset them. In 1787, a new convention met in Philadelphia; no fewer than twelve states sent delegates (only Rhode Island held aloof).

Exceeding their instructions and driven chiefly by Madison, the convention decided to work out a constitution, a basic law that was to establish a federal government, one in which states were to retain power over matters that concerned themselves alone while a central government was to have power over matters that concerned more than one state. This Constitutional Convention worked out the constitution that has successfully governed the United States for two centuries now. It was voted on in each state, and only when nine states (more than two-thirds of the number) voted to adopt the constitution did it come into effect. (Rhode Island didn't accept it till 1790.)

The Constitution unified the states and made the United States a true nation. To do so, the states voluntarily (though in some cases grudgingly) gave up some of their sovereign powers and transferred them to the central government.

Of course, since we live in a real and imperfect world and not in an idealized Heaven, the compromise that resulted in the Constitution became something to squabble over. There was steady and continuing disagreement as to just how many powers had been handed over to the Federal government. There were always those who wanted to keep those powers to a minimum and leave the states as powerful as possible. These "States' rights" men have been a force in the United States to the present time. Extreme "States' righters" maintained that since the states had voluntarily handed over some rights, they could voluntarily take them back too: they said their states could even secede from the Union.

The issue came to a head in 1860. The southern states differed from the northern states in many ways, but their differing attitudes toward Black slavery was the most emotional and divisive. In the end eleven of the southern states seceded, despite the fact that the Federal government, now under Abraham Lincoln, insisted that all the states had given up their rights permanently and could not take them back. A bloody four-year long civil war was fought, and the Federal government won. The Union remained intact.

We have the example of the United States in front of us. Can the nations of the world follow suit and form a Federal World government, with each nation controlling those affairs concerning itself only and the Federal World government controlling those affairs that concern more than one nation? What choice do we have but to try?

Naturally, it won't work perfectly from the start. The American Constitution did not. There would bound to be quarrels over the powers of the Federal World government and cries of "Foul" whenever a strongly felt issue went against one region of the Earth and for another. That happened in the United States, too, even to the extent of civil war. However, we can't afford a civil war on the world level between super-regions for the same reason we can't afford an international war right now between super-powers. It would destroy the planet.

How can we make a Federal World government work then? How can we reconcile the irreconcilables?

History may seem to make such a task an unlikely one to fulfill. The Francophones of Quebec are still fighting the English victory of two and one-half centuries ago; the Catholics of Northern Ireland are still fighting the English victory of three centuries ago; the Israelis are defending a land they last defended nineteen centuries ago. Nothing seems forgettable. And yet there is an example of the contrary also.

In the American Civil War, the southern states were defeated and humiliated after heroic and resolute battle. The "Lost Cause" is not forgotten in the South, but the bitterness is gone. There are no Confederates still fighting for independence by guerilla war and terrorism. What happened?

In the decades following the Civil War, the western territories of the United States were opened and converted into states. People flooded westward from north and south alike, from the victors and from the defeated. In the great common task of building a greater nation, the petty passions of the past, if not forgotten, were at least seen in better perspective and lost their importance.

Can *we*, then, in similar fashion, find a great project in which the nations of the world—the Soviet Union and the United States, in particular—can learn to disregard their differences in similar fashion?

It would help if we could find something that all nations held in common. To be sure, we hold our biology in common. All of us, wherever we live, however we look, whatever our culture, form a single species, entirely capable of interbreeding and of understanding each other intellectually. We have all developed language, art, and religion.

However, the superficial differences of appearance of skin, eyes, hair, noses, height, and so on, are so noticeable that the common biology is all too often forgotten, even though the similarities among us are enormously greater than the differences.

Differences in culture, though even more superficial (after all, an American-born Chinese child can become completely American in all aspects of culture, and a Chinese-born American child can become completely Chinese) produce even more hatred and suspicion. Differences in language and religion, in eating habits, in superstitions and customs, in notions of family, and in the subtle manner of social intercourse, all seem to set us irreconcilably apart.

And yet there is one aspect of humanity which is identical everywhere—science and technology.

Modern science may have been developed in Western Europe, but it has spread through the whole world. There is no such thing as Japanese science or Nigerian science or Paraguayan science, each based on a different set of axioms, following distinctive laws of logic, taking advantage of mutually exclusive laws of nature, and producing totally distinct varieties of solutions to identical problems.

The laws of thermodynamics, quantum theory, the principle of indeterminacy, the conservation laws, the concepts of relativity and biological evolution, the findings of astronomy and chemistry, are the same everywhere, and people of all nations and cultures can (and do) contribute to their further elaboration and application. Airplanes fly for all of us, television sets entertain, antibiotics cure, and nuclear bombs explode. For that matter, we all use the same resources and produce the same pollutants.

Can we not take advantage of the science and technology that humanity holds in common to find some acceptable ground on which we can all stand?

For instance, the most dramatic advance we have made in the last generation has been our advance into space. Once again, humanity has opened the possibility for a splendid increase of range, one that even surpasses the last great advance of the "Age of Exploration" four centuries ago. Human beings have stood on the Moon, and human instruments have examined at close range worlds as far away as Uranus.

The possibilities to which this extension of range can give rise are enormous. We can visualize power stations that convert sunlight into microwaves which can be beamed to Earth and converted to electricity. At its best, this can supply all the power we would need for as long as our species (or even the Earth itself) is likely to exist.

We can mine the Moon and, later, the asteroids for the material necessary to build structures in space. We can build observatories in space that can study the Universe undisturbed by our troublesome atmosphere. We can build laboratories in which to conduct experiments too dangerous for intimate contact with Earth's biosphere. We can build factories that will take advantage of the unusual properties of space (zero gravity, hard radiation, low temperatures, endless vacuum) to do things that can be done only with difficulty, or not at all, on Earth. We can transfer our industrial plant, as much as possible, into orbit and free the planet of much of our pollutants. We can even build cities in space and develop new societies.

Clearly, all of this can benefit all nations and all peoples alike. If people the world over contribute to the thought, the time, the money, and the work that will allow human expansion into the solar system, then all can claim a share in the benefits. In this huge project, the petty differences that separate us on Earth will have to shrink into relative insignificance.

There were Greeks in the fourth century B.C. who dreamed of uniting the Greek fragments in the great project of conquering Persia. There were Western Europeans in the twelfth century who dreamed of uniting the European fragments in the great project of conquering the infidels in the Holy Land. They saw war as the great unifying principle.

But we of the late twentieth century can no longer do so. It is, in any case, better, even if war were possible, to find a great unifying principle in the works of peace, as the United States did after the American Civil War—and that is what space offers us in overwhelming abundance. How criminal it would be for us not to seize that opportunity and to use space instead only as an arena for extending the threat and practice of war. The various differences among nations are not conceivably worth the destruction of civilization and humanity.

And yet to speak grandiosely of space is to risk drowning in generalities.

Where do we begin? What specific task can we undertake and make into a cooperative venture? And if world-wide cooperation is not in the cards, what type of lesser cooperation might serve as a first step?

I suppose that no one would argue the fact that by far the most dangerous rivalry in the world today is the one between the United States and the Soviet Union. Whatever destruction other rivalries may bring about, only a full-scale war between the two superpowers can bring about Armageddon.

It would therefore be in the highest degree useful to bring about some joint Soviet-American venture of note and drama, a venture which could engage the full attention of the world and to which other nations could contribute if they wished—something that could be a useful first step toward other cooperative ventures and that could generate sufficient good-will to point toward the ultimate development of a "Constitutional Convention" that can formalize a bearable global unity that preserves the distinctive customs and habits of individual nations while eliminating the domination of some nations by others.

The Moon is reachable and has been reached, and Mars is the next nearest world on which human beings could stand. (Venus, which is somewhat closer, has a temperature and an atmospheric density that cannot, at the present level of technology, conceivably be endured by astronauts.)

To send a human expedition to Mars is, at present, something that is at the extreme range of human capability. If we can carry it off, however, the knowledge gained could be extraordinarily useful, for Mars is the most Earth-like planet in the solar system, and is yet significantly different. Quite apart from the usefulness of the project, the drama should catch the emotions and the imagination of humanity.

Such an expedition is something both the United States and the Soviet Union would like to carry through, one may be sure, but the expense and effort would be all but prohibitive for either nation. It at once becomes more practical if the two nations pool their knowledge and their effort—and what a dramatic example of cooperation that would offer.

It would surely not, in itself, overcome the intransigence imposed upon us all by selective historical memory and by skewed human thought, but it would point the way—it would be a beginning.

And the way it would point—the beginning it would make—would involve human survival, and that is certainly worth the effort.

2

THE SCIENTIST
AS UNBELIEVER

If we are to think of a scientist as a possible "unbeliever," then we must assume that there is something he does not "believe." What might that be?

In the usual connotation of the word, that which a scientist is to believe or not believe are the tenets of religion, in particular those of the Judeo-Christian religious belief system of our Western tradition. (There is some question in my mind as to whether it would not be appropriate, as the world shrinks and the Middle East gains in importance, to speak of the Judeo-Christiano-Muslim religious belief system, but such a change would be purely cosmetic. It would not affect the line of argument.)

Judeo-Christian religious beliefs are heterogeneous in nature. Jews and Christians are separated by the unbridgeable gulf of Jesus Christ as Messiah. For two thousand years, Jews have refused to accept Jesus as the Messiah, Savior, and Son of God, while Christians tend to make the acceptance of Jesus as central to salvation ("Neither is their salvation in any other," Acts 4:12).

If we restrict ourselves to Christians, then there is an unbridgeable gulf, in the form of the Papacy, between Catholics and Protestants. Catholics accept the Pope as the Vicar of Christ and the head of the Universal Church, while Protestants do not. Within the realms of Judaism, Catholicism, and Protestantism there are other divisions, large and small, over matters of doctrine, which at some periods of history (such as our own) are glossed over, and which at other periods have led to fiery and explosive disputes—as fiery as the stake and as explosive as gunpowder. For over a century, from 1522 to 1648, Europe suffered under "the wars of religion," which were bloodier and more merciless than anything it was to experience prior to our own century.

Naturally, then, it makes no sense to ask whether scientists believe in (or do not believe in) baptism by total immersion, in the use of incense during church services, in the inspiration of Joseph Smith, or even in the divinity of Jesus, since many non-scientists and many sincerely and totally religious people devoutly believe or disbelieve these points or any of a thousand others.

We must find something that all segments of Western Judeo-Christianity agree upon. It is not very difficult to decide that the fundamental belief system rests in the Bible—the Old Testament for Jews and Christians alike, and the New Testament as well for Christians alone.

No matter how the various sects and divisions within Judaism and Christianity differ in doctrine, all agree in taking the Bible seriously. Different people may be worlds apart in their interpretations of the Bible and the deductions they make from it, but all "believe" the Bible, more or less.

In considering the Bible, it is important to realize that it is not all of a piece. It is not written by one person, so it is not presented from a single point of view or from a completely self-consistent picture of the world. Those who "believe" the Bible are apt to consider it as the inspired word of God, so that whether one person or many persons held the pen or pens that wrote the words, only one Mind was behind it. However, if we look at the Bible *as though* it were purely a human book, it would *seem* to be a heterogeneous mass of material.

The Old Testament consists of solid history, as 1 Samuel, 2 Samuel, 1 Kings, and 2 Kings. It also contains legendary material in Genesis and Exodus, liturgical directions in Exodus, Leviticus, Numbers, and Deuteronomy, comments on (then) current affairs, poetry, philosophy, visions of the future, and so on. The New Testament consists of four short biographies of Jesus, historical material in Acts, a series of letters on doctrinal matters, and an elaborate vision of the future in Revelation.

For much of this the question of disbelief does not arise. The histories of the Kings of Israel and Judah might conceivably be wrong in detail, but so might the history of the presidents of the United States. No one seriously questions that the historical sections of the Bible (however they might be slanted in favor of the "home team," as is true of nearly all histories) are correct in essence. Nor can anyone quarrel with the liturgical material, which is authentic by definition, just as the ritual of a Masonic society might be; or with the poetry, which is widely recognized as among the most glorious in the world.

Where in the Bible, then, does the question of belief or unbelief arise?

To answer that question, let us shift to a consideration of science. Science, properly speaking, is not a noun but a verb, not a thing but a process, not a specific set of conclusions but a way of looking at the Universe.

Science is based on certain assumptions, as all things must be. It assumes that the Universe "makes sense," that it can be reasoned about, that the rules of logic hold. It assumes that by reasoning from sense impressions one can work out the general rules ("laws of nature") that seem to govern the behavior of the Universe, that these general rules can be grasped by the human mind, and that they can be tested by experiment.

Under these circumstances, science deals with those aspects of the Universe that can be observed by the senses, that can be measured by instruments in a reproducible manner with results that do not deviate erratically from time to time or place to place or experimenter to experimenter. The Scientific Universe is by no means coterminous (at least, not yet) with the Total Universe.

Naturally, science is a cumulative process. As more and more observations are made, and more and more experiments are conducted, a broader, deeper, and more useful understanding of the Scientific Universe is obtained. It follows that we understand many aspects of the Universe far more thoroughly than our predecessors did twenty-five centuries ago, or one century ago, or, in some ways, ten years ago.

Scientists accept contemporary conclusions, not blindly, not without understanding that they are possibly temporary as well as contemporary, and not without considerable discussion and dispute, but it is a rare scientist indeed who does not accept the fact that the conclusions of today are in general more nearly correct and useful than those of a century ago, and certainly more nearly so than those of twenty-five centuries ago.

I choose the period of "twenty-five centuries ago" deliberately since much of the Bible (although in part based on older material) received its more or less present form at the time of the Babylonian Captivity of the Jews in the sixth century B.C.

It follows, therefore, that the Bible contains some sections that detail matters concerning the Scientific Universe as they were understood twenty-five centuries ago.

There is nothing essentially wrong with this. It was the best that the compilers of the Bible could do. The first chapter of the Book of Genesis, for instance, which describes the manner of the creation of the Universe, accepts the cosmogony worked out by the Babylonians, which at the time was the most advanced in the world. The Biblical version improved on the Babylonians, eliminating polytheistic notions, and introducing a lofty abstraction that is much more attuned to modern feelings. It might fairly be maintained that the Biblical tale of the creation is the most reasonable to have been produced by the human mind before the rise of modern science.

Nevertheless science has advanced in twenty-five centuries, and scientists do not, in general, accept the version of the creation presented in Genesis 1. Nor do they accept many of the other aspects of the picture of the Scientific Universe presented in the Bible.

To present a few examples, there is no convincing evidence of a world-wide Flood (as distinct from local flooding of the Tigris-Euphrates valley) in the third millennium B.C. There is no convincing evidence that humanity spoke a single language in that same millennium, or that that language was Hebrew.

There is no real evidence, outside the Bible, that such individuals as Adam, Eve, Cain, Abel, Nimrod—or even Abraham, Isaac, and Jacob—ever lived. The various "miracles" of the Old and New Testament, in which divine intervention suspended or subverted the operation of the laws of nature, are not acceptable as part of the Scientific Universe, and there is no evidence outside the Bible that any of them actually took place.

If we subtract all these things from the Bible; if we subtract such things as talking serpents and talking asses, of the Red Sea parting and the Sun standing still, of the spirit of the dead Samuel being raised by the witch of Endor, of angelic messages and demonic possession, there yet remains a great deal of prose and poetry concerning which there is no possibility of conflict with science.

It is therefore possible for the most rigid scientist to accept the Bible, almost all of it, without reservation. He can accept the history, the poetry, the ethical teachings.

In fact, he can even accept those portions which do not fit the scheme of the Scientific Universe, if he is willing to accept them as allegorical or figurative statements. There is no great difficulty in supposing that God's initial command, "Let there be light," symbolized the initial existence of energy out of which all else followed. Indeed, it is possible to see an equation between that command and the "big bang," which is the current scientific image of the origin of the Universe. With a little ingenuity almost any of the miracles of the Bible can be given poetic or allegoric meaning.

In this way, any scientist can, if he wishes, find no conflict between religion and science. He can work in his laboratory through the week and go to church on Sunday and experience no difficulty in doing so. It is possible for priests, ministers, and rabbis to fulfill all their religious duties and, when those are done, turn to scientific labors without a qualm.

It is not surprising, then, that many first-rank scientists have been sincerely and honestly religious. There was Edward Morley of the Michelson-Morley experiment, for instance, and Robert Millikan, Abbé Georges Lemâitre, and many others.

The catch is this, however. No matter how religious a scientist may be, he cannot abandon the scientific view in his day-to-day work and explain some puzzling observation by supposing divine intervention. No scientist of any standing has ever done this, nor is one likely to. Again, when a literal reading of the Bible is apt to promote a view incompatible with the Scientific Universe, any scientist, however religious, is bound to accept that view in some allegorical fashion.

It is not much of a catch. Not only do scientists, generally, find no difficulty in living up to this requirement, but large numbers of religious leaders and followers, who are not themselves scientists, have no difficulty in accepting the Scientific Universe and the current scientific conclusions concerning it. Specifically, they find it easy to accept a Universe that is billions of years old in which our Earth and its load of life (including *Homo sapiens*) developed by extremely slow stages.

We can, however, begin from the other end. Suppose we assume that the Bible is divinely inspired, that every word represents ultimate truth and is inerrant. This is the Fundamentalist view.

In that case, the Bible represents, once and for all, a view of the Universe which cannot be changed. The only function science can have in such a Fundamentalist Universe is to uncover evidence which supports that Universe or, at the very least, does not conflict with it. What happens if science uncovers evidence that does conflict with the literal word of the Bible? According to the Fundamentalist view, that evidence is wrong or is, at the very least, misleading.

In the Fundamentalist view, then, the Universe—all of it—was created some six (or possibly ten) thousand years ago in a period of six days. The Earth was created first, then the Sun and Moon and stars. Plant life was created before the Sun was, and every species of plant and animals was created separately in such a way that one could not evolve into another form.

The firmly established scientific view that the Universe is probably fifteen billion years old, that it began in a gigantic explosion out of which the galaxies slowly formed, that the Sun and its attendant planets were formed perhaps ten billion years *after* the Universe was born, that Earth is nearly five billion years old, that human-like beings slowly developed out of previous non-human forms of life a few million years ago, and that *Homo sapiens* established itself on Earth hundreds of thousands of years ago, is entirely rejected by the Fundamentalists.

There are many other places where the Fundamentalist Universe and the Scientific Universe are totally at odds, but nothing new would be added by listing them all. The actual quarrel today has the manner of origin at its core. Was it long ago (Scientific) or recently (Fundamentalist)? Was it slowly through evolutionary processes (Scientific) or suddenly through divine creation (Fundamentalist)?

No compromise would seem possible. Scientists cannot give up the results of scientific observation and experiment or the laws of logic and remain scientists. And Fundamentalists cannot give up the literal interpretation of every word in the Bible and remain Fundamentalists.

The scientific view is that Fundamentalism is a form of religion that clings to Babylonian science of twenty-five centuries ago and by modern standards is nothing but superstition.

Fundamentalists, on the other hand, wish to appropriate the respect given to science and have their views taught in the public schools as "science." This cannot constitutionally be done if their views are backed by Biblical evidence and nothing more, since that would violate the principle of separation of church and state. Therefore, they present their Fundamentalist views without mention of the Bible, surround them with a potpourri of undigested scientific terminology, and call it "scientific creationism." However, calling a horse a golden-haired princess doesn't make it any less a horse, and "scientific creationism" is Fundamentalism.

Now then, if by religion we mean Fundamentalism (but please note that *only* Fundamentalists make that equation), and if a "believer" is defined as one who accepts Fundamentalism, and an "unbeliever" as one who rejects it—then every scientist is, by definition, an unbeliever. No one who accepts the Scientific Universe can possibly accept the Fundamentalist Universe; the two are incompatible right down to their basic assumptions.

To be sure, it is possible for people to call themselves scientists, to own pieces of paper tht give them the legal right to put initials after their name, and yet to profess a belief in the Fundamentalist Universe. Anyone can *call* himself a scientist. But, I assure you, it takes more than a piece of paper to *make* one.

—— 3 ——

THE CHOKING GRIP

There are always those who wish to clamp down on the sexy aspects of television and to return it to the purity of "Ozzie and Harriet."

Lift those necklines! Thicken those brassieres and rivet them firmly in place! Unleer those eyes! Unwiggle those hips! And watch what you say!

The penalty? The New Puritans will reckon up the sex-points, watching narrowly for every hint of cleavage, for every parting of kissing lips, and will then lower the boom on all advertisers who fail the test. By striking at the pocket-nerve, they hope to produce a new kind of television that will be as clean, as smooth, and as plump as eunuchs usually are.

In a way, I sympathize with the purity brigade. The emphasis on top-front and bottom-rear gets wearisome after a while and blunts their very real values through over-familiarity. Sniggering isn't the most amusing form of laughter and eye-brow-wiggle-cum-leer doesn't make for the most pleasant countenance.

Nevertheless, there is nothing that salaciousness can do that, in the long run, will do as much harm as will the choking grip of orthodoxy. Set up a standard of purity and right-thinking and begin to demand that this be the standard for all; take to watching your neighbor lest he disagree with your conception of right and punish him if he does so; make certain we all think alike or, if necessary, don't think at all; and there will be a general crushing of intellectual liveliness and growth.

In a healthy society, dissent is free and can be endured. If some of the dissent is unpleasant to many people—that's what makes it "dissent," and that's the price you pay for social health.

If you think that a sex-ridden society, or a permissive society, or a think-as-you-please society is *not* healthy, you have but to try the kind of society in which unbridled repression sees to it that you think, write, say, and do only what some dominating force says you may, and you will then find out what an unhealthy society really is.

It's happened before on many occasions and there's no mistaking the results.

In the sixteenth century, the Protestant movement split the Church in Western Europe. In some regions the result was a move in the direction of a religiously permissive society to a greater or lesser degree—in the Netherlands, in Great Britain, even just a bit in France.

In other regions, orthodoxy gathered its forces and clamped down, determined that opposition views would be excluded totally. Nowhere was the drive for orthodoxy more thoroughgoing, nowhere did it make as much headway, as in Spain, which was then the most powerful nation in Europe.

But that power faded as the grip of orthodoxy closed about the national throat. There were many reasons for the decline, but surely one of them was the fear of thinking that permeated the land. Spain did not share in the fermenting growth of commerce, manufacturing, science, and technology that boiled and bubbled in France, Great Britain, and the Netherlands, each of which beat Spain and, for a time, inherited the world.

Italy was a divided country in early modern times and had no political or military power to speak of, but it had a vigorous intellectual life throughout the Middle Ages, and art, science, and literature reached a glittering peak between 1300 and 1600.

The greatest of all Italian scientists was Galileo, who in his old age was brought before the Inquisition in 1632 for espousing too effectively a variety of heretical opinions. He himself was treated gently enough. He was not actually tortured and he lived out his last years in peace. It was Italy that suffered, for the cold water dashed on intellectual venturesomeness in the name of orthodoxy played its part in the land's scientific decline. Italian science never sank to the almost zero-level of Spain, but leadership was gone. That passed to the northern nations.

In 1685 a powerful French king, Louis XIV, put an end to a century-old toleration of Protestant Huguenots. Many thousands of these Huguenots were driven out of the land so that France might become solidly orthodox. Those Huguenots fled to Great Britain, the Netherlands, and Prussia, subtracting their intellectual and technological strengths from France, and adding them to France's enemies. That was surely a contributing factor to France's slow decline thereafter.

In nineteenth century Germany science had its greatest flowering up to that point. Germany, as a whole, was not remarkable for its liberty, but, like Italy, it was split into numerous states. Unlike Italy, however, there was no religious uniformity among those states, and what one state did not allow another would. In the competition among them, dozens of great universities rose and flourished. Intellectual advance expanded marvellously.

In Austria, however, Germany's largest state and therefore the most effectively orthodox, science and technology advanced least, and progress of all sorts languished.

Germany showed the reverse picture in the twentieth century. The most madly repressive regime modern Europe had seen was established by Hitler in 1933. In the space of a few years, he clamped a ferocious Nazi orthodoxy first upon Germany and then upon much of Europe. Jewish scientists and, indeed, scientists of any kind who were insufficiently Nazi were harried, driven out. These refugees subtracted their strength from Germany and added it to Great Britain and the United States, thus contributing to the defeat of Germany.

Soviet and Chinese insistence on Communist orthodoxy has undoubtedly weakened both nations. How can it have helped but do so? Any national practice that deadens the spirit and suppresses variety and activity of thought will surely produce a national coma.

Are there examples of the opposite? The one most frequently cited is the sexual abandon of the Roman Empire—depicted so well in "I, Claudius"—which supposedly led to the fall of Rome.

Not so. That sexual abandon was confined to a small aristocracy and was greatly exaggerated by the scandal-mongers of the age for political reasons. In any case, under those wicked early Emperors, the Roman Empire was at the height of its power. It was in later centuries, when pagan orgies vanished, but when the Empire lay under the power of a religious orthodoxy that condemned all dissent, that it fell.

So perhaps we had better endure the naughtiness of television and hope that good taste among the viewers will prevail in the long run. To set our foot on the deadly path of censorship and suppression will surely lead to destruction in the end.

Afterword: This essay was written in 1981, and I'm glad to say that in the years since the move for "purity" in television has not notably advanced. I have always been ruefully amazed, however, that even in the United States there seem to be so many who don't know what the phrase "freedom of expression" means.

After this essay first appeared, I received a number of letters from people who felt that the full majesty of the state should be brought to bear upon those who had the nerve to disagree with the letter-writers' views about sex. They were themselves ready to be the censors. Anything that offended them offended the Universe (or God, perhaps). They had all the answers. They knew.

Naturally, I am willing to let them have their say. I just wish they were willing to let me have mine.

Oh, some of the letters objected to my remark about the program featuring Ozzie Nelson and Harriet Hilliard. I had no real hatred for the program. I frequently listened to it on the radio or watched it on television a generation ago. It's just that it was the blandest of the bland and was about as tasty as a slice of store-bought white bread.

4

HUMAN MUTATIONS

Human beings in their past history must have had mutations, sudden changes in their structure, because of changes in their genes. Such things must be happening all the time. Perhaps every one of us has one or more mutations, a gene that changed in the body of one of our parents and was inherited by us.

Such mutations can be very small and pass unnoticed. Sometimes, however, they can be very noticeable, and then they are usually harmful. Once in a while, though, a mutation might be beneficial, produce something useful. Such mutations result in a human being who is better off, more capable, who may live longer and have more children, so that the mutation is passed on to an increased number of individuals. Each of these passes it on, too, to more and more, until, in the end, almost all human beings have it. The mutation becomes part of an evolving humanity through "natural selection."

What are the most important and useful mutations that have taken place in the past? What are the mutations that have made human beings what they are today?

The key mutation took place four million years ago or more, when a certain ape-like creature was born with a spine that curved in such a way that it was easier for him to stand up on his hind legs and balance on them for a long period of time.

Modern apes can stand on their hind legs when they want to. So can bears and some other animals. All of them, however, have a spine that curves like a shallow arch. Such spines are not well-adapted to being tipped on end. It's an effort for animals with them to stand on two legs, and they quit after a while. A human spine isn't a simple arch, however. It is S-shaped. We can stand upright for hours at a time balancing on that spine. We can walk and run on two legs with ease. It's not a perfect mutation, for it puts so much strain on the spine that many people develop pains in their lower back as they grow older. But on the whole it serves us well.

We don't know how this mutation came about because we weren't there to watch, and only very few human bones have survived from all those millions of years ago. It probably happened in steps. Every time someone was born with a spine that made it easier to stand on two legs, he was better off. Possibly that was because it raised his eyes higher and he could see food, and enemies, at greater distances. Each of these mutations spread by natural selection, and finally there were small ape-like creatures that walked upright as well as you and I.

It might seem to you that this is a mutation that affected only the physical body, but strangely enough it didn't. Everything we think of as human may have come about because of this ability to stand on two legs.

Our eyes being higher, they were used more to see long-distance, so more information continually flooded into the brain. What's more, the forelimbs, which human beings no longer used for standing on, were free to hold things, pick them up, manipulate them, feel them, carry them to the mouth or eyes. Again, a great deal of new information flooded into the brain.

As a result, if any new mutation took place that made the brain larger or more efficient, the result was very useful to the two-legged creature, for the brain could then handle the flood of information more easily. Such a mutation would, therefore, spread rapidly by natural selection. The brain, under those conditions, would (and did) increase in size. Over the last half-million years, for instance, the human brain just about tripled in size until it is now *enormous* for a creature no larger than ourselves.

In this way, a mutation that seemed to be a purely physical one resulted in further physical mutations that ended in making us the most intelligent creature that has ever existed on Earth. Only we learned how to develop speech and make use of fire, and then work out an advanced science and technology.

And what now? Can we expect further mutations?

Of course! They happen all the time, as I said earlier. I wonder, though, if further physical mutations are likely to be incorporated into human structure by natural selection very often. After all, our physical bodies are not the important thing about us anymore. Our bodies can't fly as birds can, for instance, but who cares? We have airplanes that can go faster than any bird, and we even have rockets that will take us to the moon, and no bird can fly there.

No, what is important about human beings is their enormous brain. What if there were mutations that affected it; small changes that could improve the efficiency of the brain or give it new powers? We don't have any record of such changes, at least no changes that are big enough to notice, but then—

Perhaps we don't look carefully. Or perhaps we don't quite understand what we see. Perhaps people with changed brains simply seem weird to us. Perhaps —

5

THE HOLLOW EARTH

In a way, I am glad that the mail I get still has the capacity to surprise me with its manifold evidences of crackpottery. Just as I am ready to sink back into the sad certainty that I have seen it all, and that nothing more can break life's round of dullness, something astonishing comes along and socks me right in the funny bone.

Let me begin by saying that any number of people are perpetually publishing books at their own expense in which they put up for public gaze their pearls of nitwittery. This, in a way, is good, since it demonstrates our belief in the freedom of speech and press. If, in the name of sanity and good sense, we were to be lured into suppressing these examples of limping lunacy, we would be setting a precedent for quashing the expression of any view that someone or other dislikes and you and I would surely be in trouble (see Essay 3, "The Choking Grip").

Of the various examples of published pinheadedness of which I here speak, a sizable fraction seems to reach me, since, for some reason or other, every inventor of such vacuosities is firmly convinced that I am so broad-minded, and so open to the unusual, that I would be sure to welcome junk with loud yelps and cries of approval.

Naturally, I don't. I can tell nonsense from sense and distinguish the mad from the unusual. I am not afraid of the merely offbeat, as I showed when I wrote an introduction to "The Jupiter Effect," but I can draw the line.

For instance, I draw the line at all notions that we live on a hollow Earth, that within the curved surface of our globe is a vast cavern with a central radioactive "sun," that on this inner surface there is a world like our own. These notions belong in the same category as the belief in elves, leprechauns, and creationism.

Imagine, then, my astonishment at receiving a little soft-covered book that not only propounds the hollow Earth belief, but in great huge letters on the cover takes *my* name in vain as another proponent. Within the covers it quotes passages from various books of mine, taken quite out of context, and then interprets them with non-sequiturish madness as supporting the hollow Earth.

For a moment, I considered a possible lawsuit, since aspersions were being cast on my competence as a science writer, and I was being damaged in consequence.

Common sense quickly prevailed, however. The author undoubtedly had no money I could seize, and legal expenses would be high. Furthermore, money would in no way compensate for the damage, if there was any; and I wasn't likely, in any case, to demonstrate damage, only hurt feelings, for it was clear, on the face of it, that no rational person could possibly take the monumental lunacy of the book seriously. And then, any legal action would merely publicize this lump of unreason and do the author a favor.

But where did the notion of the hollow Earth come from?

To begin with, everyone assumed the Earth was flat, and it was indisputable that caves existed. For the most part caves were unexplored and no one knew how deeply they penetrated. It seemed sensible to suppose that the Earth was a thin, flat slab with as much empty space beneath it as there was above it.

The existence of volcanoes rather indicated that the space beneath was not very comfortable. In the Greek myths, giants who rebelled against Zeus were chained underground and it was their writhings that caused earthquakes, while underground forges of divine or demonic smiths occasionally overflowed and produced volcanoes.

The Old Testament, in places, seems to preach that good is rewarded and evil is punished right here on Earth, but this didn't hold up. The evidence to the contrary was so overwhelming that even as inspired writ the idea couldn't be accepted. One had to suppose that the rewards and punishments came after death and, being unwitnessed, they could be made all the vaster—infinite bliss for the people you like and infinite torment for the people you dislike. The bliss was in heaven above, and the torment in hell below. Volcanoes made it seem likely that hell was a place of fire and brimstone.

Even after the Greek philosophers, particularly Aristotle, demonstrated Earth to be a sphere, that sphere had to be considered hollow, since otherwise there was no place for hell. Dante's detailed description of the Earth had it the hollow spherical center of the Universe with a series of concentric heavens surrounding it and a series of concentric circles of hell within it. The three great traitors, Judas, Brutus, and Cassius (one for God, two for Caesar) were frozen into the ice at Earth's very center.

Thus, the notion of the hollow Earth was made virtually part of religion, and when modern science was born, scientists struggled to make it sensible as well as pious. In 1665 the German scholar and Catholic priest Athanasius Kircher (a first-class scientist who speculated on the possibility that newly discovered microorganisms might be the cause of disease) published "Subterranean World," the most highly regarded geology text of its time. It described the Earth as riddled with caverns and tunnels in which dragons lived.

Naturally, the thought of another world just a few miles under the surface on which we live was infinitely attractive to science fiction writers and, after Kircher's time, there was a steady drumbeat of stores about the Earth's hollow interior. The best of them was Jules Verne's "A Journey to the Center of the Earth," published in 1864. Verne described underground oceans and dinosaurs and brutish man-apes down there. The most popular hollow Earth stories were, perhaps, those by Edgar Rice Burroughs about "Pellucidar," the name he gave the inner world. The first of those books was published in 1913.

The hollow Earth was by no means left to the domain of science fiction, however. In the 1820s an American named John Cleve Symmes insisted that Earth consisted of a whole series of concentric globes with space between. These inner worlds could be reached by holes that passed through them all and penetrated the outermost surface on which we live. And where were these holes on our world's surface? Where else but at the North Pole and South Pole, which in Symmes's time had not yet been reached and could therefore be safely dealt with?

Symmes presented no real evidence in favor of his hypothesis, but he gathered around himself the kind of devotees that need no evidence, the kind that are made the most furiously devoted to nonsense by any attempt to demonstrate its irrational character. They were people of the same sad sort that cling ferociously to the Velikovskys and von Danikens of our own day.

But is it possible that the Earth *is* hollow?

No, it isn't. The evidence showing the Earth to be a thoroughly solid body goes back before Burroughs, before Verne, and before even Symmes. From his first lecture on the subject, Symmes was talking through his hat.

Why? Because in 1798 the English scientist Henry Cavendish determined the mass of the Earth, his preliminary figure being almost exactly that given by today's most delicate measurements. This mass showed Earth's average density to be 5.5 grams per cubic centimeter.

The density of the ocean is roughly one gram per cubic centimeter, that of the surface rocks an average of 2.8 grams per cubic centimeter. In order to raise the overall average to 5.5 grams per cubic centimeter, the lower regions of the Earth must be considerably denser than the surface rocks. It is all we can do to account for the overall density of the Earth if we suppose it not only to be solid, but to have an interior core of nickel-iron under pressure. To imagine that the overall density can be reached if the Earth were hollow is simply out of the question.

But that's not all. We can calculate how the temperature and pressure of the rocks must increase with depth, based on observations in mines and on theoretical considerations. These calculations are not precise and somewhat different figures are reached by different scientists, but all the figures agree in this: At a not too considerable distance below the surface, the rocks become hot enough to be plastic under the pressures they are subjected to. In other words, if you suddenly created hollows deep underground, the surrounding rock would be squeezed into those hollows, which would fill up and disappear in short order. Thus, not only is the Earth not hollow, it can't even be made to be hollow.

Finally, seismologists have been studying earthquake waves for years. From the manner in which those waves travel through the Earth, from their speeds, from their changing directions, it is possible to work out many details about the interior—the exact distance underground where hot rock changes into molten metal, for instance. And all the studies agree on one thing—no hollows!

So the hollow Earth doctrine is a potpourri of poopery, and an inanity of insanity. So say I!

6

POISON!

Life is the end result of thousands of interlocking chemical reactions, most of them controlled by specific enzymes. The enzymes are "catalysts," substances capable of speeding chemical reactions without themselves being changed in the process.

Enzymes do their work by offering a surface that is just suited for the proper positioning of the molecules undergoing the reaction. The enzymes are protein molecules built up of hundreds or even thousands of atoms, and each is so designed that it will catalyze only *one particular* reaction. There is a separate enzyme for each reaction, and by modifying the enzymes—activating, deactivating, reactivating them—the overall nature of the reactions characteristic of life are controlled.

It is because of the differences in enzyme content that a liver cell is different from a muscle cell or a brain cell. It is because of the differences in enzyme content that the trillions of cells making up a human being produce an organism different from the trillions of cells making up a giraffe or an ostrich—or a tree.

Any substance that will stick to an enzyme and distort its surface so that the enzyme can no longer perform its function will, obviously, disrupt the working of the cells of an organism. The distorting substance is a "poison." Since enzymes are present in tiny quantities, it takes but a small amount of poison to knock out enough of them to produce serious illness, or even death.

A number of organisms, in the course of evolution, have developed the ability to form poisons of one sort or another either in attack or defense. Where attack is concerned, snakes, scorpions, and toads have their venoms: bacteria produce their toxins. In defense, plants which are eaten by myriads of animal organisms from insects to human beings, and must endure this helplessly, sometimes develop complex molecules that are most efficient poisons. Animals evolve the ability to avoid these poisons by finding them noxious to the taste—those that don't, die. The result is that poisonous plants avoid being eaten, at least to some degree.

Human beings in prehistoric times, driven by hunger or curiosity, are bound to have tasted everything. They undoubtedly encountered items that, when eaten, made them drunk, or produced hallucinations (a mild form of poisoning), and such things were sometimes enjoyed and sought out deliberately. On the other hand, there were also some items that killed. From desperate individual disaster, people learned to stay away from certain mushrooms, for instance, or berries, or leaves.

On the other hand, it must have occurred to human beings on many separate occasions that poisons could be useful. A poisonous plant might be mashed up and an arrow point might be smeared with the mess which might then be allowed to dry there. If an enemy is wounded by such an arrow, he might die even if the wound is a superficial one. This was such a convenient practice that the word "toxic" comes from the Greek word "toxon," meaning "arrow."

Death in organized conflict tends to be accepted as a sad visitation by those suffering the casualties; those who inflict the casualties can be seen as heroic. But what if someone's death is desired, and achieved, over a *private* quarrel? Then it is considered murder, and any reasonably advanced society views that with sufficient concern to want to punish the murderer.

Any sensible murderer would, therefore, find himself compelled to devise a plan of murder that would enable him to avoid punishment. He would lay in ambush for his victim, and then, in secret, smash him with a blunt instrument or stick him with a knife. He would then sneak away and hope that no one would find out who did it. However, the murderer usually has a motive that is known to the community, and he will be suspected even if the murder is unwitnessed. It would be better, then, if the death is not seen to be murder in the first place.

Suppose a poison mushroom is chopped up fine and added to otherwise harmless food that the victim intends to eat. The victim will enjoy his meal and then, sometime afterward, when the murderer is nowhere in the vicinity, he will die. There will be no obvious sign that murder was done; no cut, no smash, no blood, no break. It will be the kind of death that might have resulted from disease or from some internal stroke or seizure. Prior to the days of modern medicine, there were numerous fatal diseases that weren't understood at all, and who was to differentiate between such a disease and deliberately administered poison?

For that reason, poison became a favorite means of killing, and it was so common that the situation was often reversed. Instead of poison being considered mere disease, disease was often considered deliberate poisoning. Right up through modern times, any public figure who died suddenly, especially while not yet aged, was supposed to have been poisoned by his enemies. Even if he died as the result of a lingering disease, he was frequently supposed to have suffered the result of slow poisoning.

So while history has its poisonings, the amount was rather exaggerated than otherwise. Perhaps that is why we have the history of the Borgias (a famous Spanish-Italian family, one of whom was Pope Alexander VI) who were popularly supposed to have poisoned half the people who dined with them. In 1679 there was a sensational investigation of a murder-by-poison organization in Paris which was apparently patronized by important people, including Madame de Montespan, mistress of King Louis XIV.

Nowadays, however, thanks to modern science, things have changed considerably. Medical pathologists know how to detect traces of poisons in such a delicate and unmistakable manner that it is very unlikely that a victim of poisoning is going to be mistaken for a victim of disease if a careful autopsy is performed. On the other hand, poisons far more deadly than those available in earlier ages are now obtainable. Perhaps one-fifteen-thousandth of an ounce of botulinus toxin will be enough to kill a person.

So poisoning is still something to be considered.

——— 7 ———

COMPETITION!

The first games we know of in western literature are those described in the twenty-third book of *The Iliad*, in which the funeral rites of Achilles' friend Patroclus are described. As part of the rites, the Greek chieftains participated in games designed to show their athletic vigor.

There was a chariot race, a boxing match, a wrestling match, a foot race, a gladiatorial contest, an archery match, and a spear-throwing competition. All these things were elements of combat and it reminds one of the play of young animals, in which there is always an element of the skills that will be needed in the serious business of life. The kitten pouncing on a leaf will someday be pouncing on a mouse; the two puppies snarling and biting at each other in exuberant fun will someday be doing the same thing in earnest to establish domination, to win food, or to gain a mate.

Competition, in other words, is a deadly serious thing.

Passing out of legend and into history, it was the Greeks who made important rituals out of games, holding them periodically as part of religious festivals. The most important of these was the quadrennial competitions held at Olympia, in southwestern Greece, in honor of the supreme god, Zeus. We refer to them as the Olympic games.

According to tradition, the first Olympic games were held in 776 B.C., and they were held every four years thereafter without a break for nearly twelve centuries, until the Christian Roman Emperor Theodosius put an end to them because they were a pagan festival (which they were).

During the twelve-century interval in which they were celebrated, the Olympic games were open to contestants from every Greek-speaking city, wherever it might be located, from the Crimea to Spain. The games were, in fact, one of the three great bonds that held together the thousand independent cities of Greece. (The other two were the Greek language and the Homeric poems.) So important were the games that even wars were adjourned long enough to allow contestants to travel to and from Olympia and to compete in peace.

For fifteen centuries after Theodosius had put an end to them, the Olympic games remained a historic memory, but then they were reinstituted in 1896. Since then, they have been held every four years except when World Wars I and II were in progress. (It is a measure of the decline of civilization that nowadays the games are adjourned for war rather than vice versa.)

Ideally, in the Olympic games it is amateurs that compete; that is, contestants do it not for money, but for glory. In ancient times, the only award for winning was a crown of leaves. However, human beings are human beings and we need not think that the crown of leaves was all, just because it was supposed to be all. Winners gained imperishable glory; great poets wrote odes in their praise; they were honored in all sorts of ways; and their names were inscribed in the record books. If they did not make money directly, their status as winners made it possible for them to gain in many ways. (Nowadays, the amateur winners can make money by endorsing products, for instance. No one says they must die of starvation.)

There are two aspects of games, however, which don't figure much in the idealism with which they are surrounded. The Olympics, ancient and modern, may be hymns to amateurism and glory, but the uglier aspects of nationalism leave their mark. It is not the contestant only who wins, but the city or the nation he or she represents. In modern times, certainly, there is a constant adding up of medals for each nation, and a steady drumbeat of national pride or national resentment over winning and losing. Right now, in particular, it is considered extremely important as to whether the United States or the Soviet Union is ahead in medals. That would not be at all bad if it were a substitute for war, but it is very bad when war remains a possibility and the bad blood over sports adds to the hatreds that might spark a war.

Then, too, the members of the audience do not necessarily merely watch and approve of athletic skill and endurance. They do not even merely let themselves be influenced by irrelevant causes such as national pride. They often back their opinions, whether shrewd or nationalistic, by money and bet (in the aggregate) huge sums on the outcome.

This is especially true of professional athletic contests, where one might wonder sometimes whether there is any interest in the outcome at all, except as a matter of personal profit and loss. Is it conceivable, for instance, that crowds will watch a horse race or a football game without betting on it?

It is not surprising, then, that emotions run ridiculously high among spectators. Soccer games, the favorite spectator sport outside the United States, are sometimes bloodbaths, as spectators turn upon each other violently at some decision of the referee that fills one side with glee and the other with fury. Or, out of delight at victory or rage at defeat, spectators may turn recklessly on the city in which the contest has taken place, inflicting severe damage upon it.

One more point should be stressed. We think of the Olympic games primarily as athletic contests, but the ancient Greeks did not limit them to muscular effort at all. They considered the whole body, mind as well as muscle, to be important, and the production of tragedies and comedies, as well as the reading of literary works, were also among the contests.

Afterword: Several of the essays in this book, such as "Human Mutations" (#3), "Poison" (#6), and this one you have just read, together with some yet to come, were originally written to serve as introductions for science fiction anthologies which I helped edit and which were built around some theme. (See the list of acknowledgments.)

I include these essays because it seemed to me they were of general interest. However, it was necessary for me to end them with a rousing description of the excellence of the stories included in the anthology. Such description would have been out of place here and so I cut them out. If, then, the endings seem to be rather sudden and to lack the usual Asimovian flourish, that's why!

——— 8 ———
BENJAMIN FRANKLIN
CHANGES THE WORLD

I write numerous essays on science for young readers. This book, on the other hand, is intended for adults. I therefore debated with myself for a considerable period of time as to whether to include this essay or not. After all, you will have no trouble seeing that it was written for youngsters and this may jar you.

However, it makes an important point, something I desperately wanted young people to understand—and I wanted you older people to understand it, too, so I said "What the heck" and included it.

Besides, I'm rather pleased with the way in which I write on science for young people, and I thought I might show off a little with this essay.

The world was a cruel and frightening place to our ancestors. Nobody knew what made it work. The rain might come or not come. There might be early frosts, or floods. Almost anything might spoil the crops and bring on starvation. Disease might strike suddenly and kill domestic animals or the people themselves.

No one knew why rain should fall some times and not others, or where the heat and cold came from, or what caused sickness. The world seemed so puzzling that many people decided that it was run by demons or spirits of one sort or another.

They felt it was important to try out ways of getting on the good side of the demons, of keeping them from getting angry. Everyone worked out some kind of superstition he thought would keep bad luck from striking him.

We still have superstitions today. People think that if they knock on wood that will keep bad luck away, or if they light three cigarettes with one match that will bring bad luck. They think that if they carry a rabbit's foot that will bring good luck, and so on.

It might make people feel better to think they can keep themselves safe in all these silly ways but, of course, none of these superstitions have anything to do with the real world. None of them work.

People who study how the world really works are called scientists. They do their best, for instance, to find out what really causes disease, and how the weather really works.

Modern science began about 1600 but for about a hundred and fifty years it didn't seem to have anything to do with ordinary people. Scientists found out that the Earth went about the Sun instead of the other way around, but what difference did that make to how well the crops grew or how healthy human beings were?

Then in 1752, for the first time, science was used to save human beings from disaster, and that changed the world. For the first time people turned to science and not to superstition to keep harm away.

You might have thought this would have happened in Europe where, in those days, science was most advanced. It didn't. It happened in Philadelphia in the American colonies. In those days, Europeans thought that only ignorant farmers lived in the colonies, but they were wrong. Benjamin Franklin lived in the colonies and he was one of the cleverest men in the world.

In the early 1700s, the scientists in Europe were very interested in electricity. If they rubbed rods of glass or sealing wax, those rods attracted light objects like feathers and small bits of wood. The rubbed objects were said to be charged with electricity.

Some devices could be charged with a great deal of electricity. A particular device studied at the University of Leyden in the Netherlands was called the "Leyden jar."

If a Leyden jar is filled with a particularly large charge of electricity, that electricity might suddenly pour out the way air pours out of a punctured balloon. When electricity pours out, or "discharges," it heats the air so that there is a little spark. The air expands with the heat and then contracts again, making a sound like a little crackle.

Over in the colonies, Benjamin Franklin was interested in electricity and he experimented with Leyden jars, too.

He studied the way they discharged. If he attached a small metal rod to the Leyden jar, the discharge came off the end of the rod. If the Leyden jar was charged high enough and if something was brought near the rod, a spark would shoot off the rod and there would be a crackle.

The thinner the rod, the quicker the discharge would come. If you used a very thin rod with a sharp end, you couldn't build up a charge in the Leyden jar at all. As fast as you rubbed electricity into it, that electricity would leak out of the sharp end of the rod. It would leak out so quietly there would be no spark and no crackle.

Some people said the spark and crack were like a tiny lightning and thunder. Franklin thought of it the other way. Could real lightning and thunder be a large, large electric discharge from a cloud or from the earth?

This was an important thought because everyone was afraid of lightning. It struck without warning. It could set a house or barn on fire. It could kill an animal or a human being. The ancients thought that lightning was a weapon used by the gods. The Greeks thought that Zeus hurled the lightning. The Norsemen thought that Thor did.

If Franklin could find out that the lightning was an electric discharge, it might be possible to understand the lightning more—and fear it less.

In June 1752 Franklin made a kite and tied a metal rod to it. He ran a long twine from the kite and placed a metal key at the bottom end of the twine. He intended to fly the kite when a thunderstorm was coming up to see if electricity would flow from the clouds down to the key. He didn't hold the twine with his hand because he didn't want the electricity to flow into him and kill him. He tied a silk thread to the twine and held that because electricity doesn't travel through silk.

He flew the kite and when it vanished into a storm cloud, he carefully brought his knuckle near the key. The key discharged and produced a spark and a crackle, just as a Leyden jar would. And the spark felt the same on his knuckle as it would have if it had come from a Leyden jar.

Franklin had an uncharged Leyden jar with him. He brought it near the key and electricity flowed from the clouds into the key and from the key into the Leyden jar. The Leyden jar was charged with electricity from the sky and it behaved as though the electricity had been produced on Earth.

Franklin thought that this meant that the lightning in the sky would follow the same rules that electricity on Earth would.

During a thunderstorm, the ground could become filled with a charge of electricity. If it did, there might eventually be a huge discharge—a lightning bolt. If the discharge worked its way through a building, the heat could set the building on fire.

But Franklin had found that if a thin rod was attached to a Leyden jar, it wouldn't build up a charge. The charge would leak out of the sharp end of the rod as quickly as it was built up and there would be no spark. Suppose the same was done to a building.

Suppose a thin metal rod was placed on top of a building and connected to the ground. In a thunderstorm, the ground under the building would not build up a charge because that charge would leak quietly away through the thin rod. The house would therefore not be hit by lightning.

Franklin called such a device "a lightning rod." Every year, he published an almanac in which he included information about all sorts of things. In the 1753 edition, he described how to put a lightning rod on houses and barns to keep them from being hit by lightning.

It was such a simple thing to do, and people were so afraid of lightning that soon after the almanac came out lightning rods began to be placed on houses all over the colonies. They were used in Europe, too.

And it wasn't a superstition; it worked! For the first time in the history of mankind, one of the terrors of the world was beaten—and it was beaten by science. Never mind spells and magic; if you understood what lightning was and how electricity worked, you could take advantage of that knowledge.

Beginning in 1752, people could see for themselves that science worked and superstition didn't. In 1767, for instance, the people of the Italian city of Brescia stored a great deal of gunpowder in the cellar of a tall building which did not have a lightning rod in it. The people thought it was safe without a lightning rod because it was a church.

But the building was struck by lightning during a storm and all the gunpowder exploded. The explosion destroyed a great deal of the city and killed three thousand people.

After that, there was no argument about lightning rods.

Scientific knowledge continued to advance and to help mankind. In 1798 an English doctor learned how to use scientific knowledge to prevent smallpox from attacking people. That was the beginning of the victory of science over sickness.

In the 1840s doctors learned how to use certain chemicals to put people to sleep during operations. That was the beginning of the victory of science over pain.

In many, many ways science helped human beings where superstition had just fooled and confused them.

And it all started with an American colonial named Benjamin Franklin, who flew a kite in a thunderstorm and changed the world.

9

FIFTY YEARS OF ASTRONOMY

In October 1935 the Hayden Planetarium opened. Astronomy seemed a highly developed subject. There were huge telescopes. Distant Pluto had been discovered. The universe contained vast numbers of galaxies each as crowded with stars as our own Milky Way, and the whole of it was expanding. It seemed there was little more to expect.

As it happened, however, astronomy was entering its golden age. Fifty years of revolutionary development lay ahead. So much has taken place that here we can only sketch the major developments.

The most familiar astronomical body is, of course, the Earth itself, but in 1935 we knew surprisingly little of its geological development. In 1953, however, the Great Global Rift was discovered, and with time it came to be realized that the Earth's crust consisted of half a dozen large plates, with a number of smaller ones, all well-fitted together and all in very slow motion. It was these plates and their motions ("plate tectonics") that accounted for mountain building, island chains, earthquakes, volcanoes. It even helped explain the course of evolution.

Human beings hurled themselves off the Earth, too, something no one but a few science fiction writers took seriously in 1935. The first artificial satellite was launched in 1957. A probe went around the Moon in 1959 and sent back the first photographs of the far side, something no human being till then had ever seen. Eventually, the entire Moon was mapped in as nearly great detail as the Earth is.

In 1961 the first man was placed in orbit, and in 1969 the first men set foot on the Moon. Eventually, many pounds of moon rocks were brought back to Earth and the Moon became part of the human range.

The advent of rockets and probes advanced our knowledge of the solar system as has nothing else since the invention of the telescope.

For three-quarters of a century, it was thought that Mercury and Venus faced one side to the Sun at all times and had days as long as their years. In 1965 radar reflections from Mercury's surface showed that its day was two-thirds as long as its year. Every portion of its surface had both day and night. Similar studies of Venus showed that the length of its day was *longer* than that of its year, and that it rotated in the "wrong" direction, from east to west. Moreover, microwave radiation from Venus was detected and gave the first hint that it was much hotter than had been thought. Its surface temperature is everywhere hot enough to melt lead so that there is no liquid water on it. Its thick, unbroken cloud layer is composed of diluted sulfuric acid, and its atmosphere, 90 times the density of Earth's, is mostly carbon dioxide.

Probes have mapped most of Mercury's surface in detail, and radar observation has mapped Venus's surface rather crudely.

Probes have also studied Mars, and, in 1976, even landed on it. The long dreamed-of canals of Mars do not exist, but craters do. (Indeed, with a few exceptions, craters exist on every airless planet.) Extinct volcanoes also exist on Mars (and possibly on Venus). Its atmosphere is one percent as thick as Earth's and is mostly carbon dioxide. Its temperature is as low as that of Antarctica. Analysis of Martian soil makes it seem extremely doubtful that any life exists there.

Finally, probes have taken photographs of Jupiter and Saturn at close quarters and shown us those giants in far greater detail than could be hoped for from Earth-based observations. In a way, the surroundings of the planets were the subject of more startling discoveries than the planets themselves were.

Jupiter was surrounded by a giant magnetic field that would make human exploration extremely difficult, if not impossible, and it also possessed a thin ring of debris invisible from Earth. Ganymede and Callisto, its two outer giant satellites, were cratered and icy. Europa, the smallest of the four giants, was covered with a world-wide glacier, criss-crossed by cracks, and probably with a liquid ocean beneath its surface. Io was the great surprise. It had active volcanoes that spewed out sulfur so that the surface was red-orange in color.

The great surprise in connection with Saturn was its rings, which turned out to be extraordinarily complex, being made up of hundreds or even thousands of concentric ringlets. Dark "spokes" were present and some of the ringlets even seemed "braided." Much of the detail of the rings remains to be explained. Titan, the largest Saturnian satellite, was thought, from Earth-based observations, to have a thin, methane atmosphere, but was found to have one that was smoggy, half again as dense as Earth's, and consisting mostly of nitrogen.

Probes have not yet reached beyond Saturn, but since 1935 Earth-based studies have discovered a fifth satellite of Uranus and a second satellite of Neptune. In 1977 a large asteroid, Chiron, was discovered, with an orbit that carried it from the neighborhood of Saturn out to that of Uranus. In 1978 Pluto was found to have a satellite, named Charon, almost as large as it was itself. Pluto, in fact, turned out to be far smaller than it was thought to be when first discovered and to be considerably smaller than our Moon.

In 1935 the source of the Sun's energy was still a mystery. In 1938, however, the details of the hydrogen fusion at its center seemed to have been worked out.

In 1958 it became clear that there was a constant spray of rapid charged particles emerging from the Sun in all directions, the so-called "solar wind." Also emerging from the Sun in all directions are particles called "neutrinos" (which were only mentioned in theory in 1935 and which very few took seriously). These solar neutrinos emerge from the fusion regions at the center and could tell us a great deal about the workings of the Sun if they could be properly trapped and studied. Attempts in the 1970s and 1980s have been made, but so far only one-third of the expected number of neutrinos have been found—which may mean that we don't yet have the correct information about what goes on in the center of the Sun.

Another point of uncertainty concerning the Sun is the verification of the "Maunder minimum," which had first been suggested back in the 1890s, but which no one then took seriously. Apparently, the Sun goes through periods in which for many decades there are virtually no sunspots. This apparently affects Earth's weather, for the last Maunder minimum (1645 to 1715) seems to have coincided with a siege of unusually cold weather. We must therefore take seriously the effect upon us of a Sun far less stable and well-behaved than it was thought to be in 1935.

When we began studying the Universe outside our solar system, our only source of data was the light radiated from the stars. Although astronomers developed the use of telescopes, spectroscopes, and cameras, the basic information still came from light, even in 1935 when the Hayden Planetarium was opened.

To be sure, radio waves from the stars were discovered in 1931, but it was not till after World War II that the technique of detecting such waves was sufficiently developed to make "radio astronomy" an important branch of the science.

It was in 1969 that an entirely new kind of star was detected through radio telescopes. Rapid radio pulsations with a period in the second-range were detected and seemed to originate in what were called pulsating stars (or "pulsars"). More and more of these were detected in the sky and a consideration of the properties of those pulses led to the realization that pulsars were "neutron stars."

Neutron stars were first imagined purely out of theoretical reasoning in 1934, a year before the Hayden Planetarium was founded, but such an idea was not taken too seriously. But now they are known to exist. They are extremely compressed stars, the remnants of supernova explosions, made up of neutrons in contact, so that a star of the mass of our Sun would be condensed into an object only about 14 kilometers (8 miles) across.

An even more condensed object would be a black hole. This would be a star (or any object) that would be condensed to the point where even neutrons would smash together and break down so that the resulting matter would collapse to zero volume (a "singularity"). The existence of black holes was first suggested in 1939, but there was no chance of detecting one until satellites were designed and launched beyond the atmosphere for the purpose of picking up x-rays from space.

In 1971 an x-ray source called Cygnus X-1 was found to be an invisible object circling a massive star. From the nature of the orbit the x-ray source was itself seen to be massive and it could only be massive *and* invisible if it were a black hole.

Radio astronomy has shown that many galaxies are "active"; that is, they are unusually strong radio emitters, such emissions tending to come from various jets extending from the centers. Some galaxies actually seem to be exploding. Even quiet galaxies, such as our own, have concentrated spots at the center that emit radio wave radiation in considerable quantities. It would appear that black bodies may exist at the center of most, if not all, galaxies, and even, possibly, at the center of globular clusters.

It would seem under those conditions that the placid appearance of the night sky as seen by light alone, whether by eye or telescope, is, in a sense, a sham. By instruments that have only become available to us in the last generation, the Universe would appear to be an unimaginably violent place. It may even be that we are alive only because our Sun circles the Galactic center in an orbit that keeps it safely in the spiral arm outskirts.

Some radio sources seemed at first to be rather dim stars in our own Galaxy, and this made them unusual indeed, since the vast majority of stars did not emit detectable quantities of radio waves. It was suspected, therefore, that they were something more than stars and were called "quasi-stellar" (that is, "star-like") radio sources, or, for short, "quasars."

In 1963 it was discovered that these quasars were far more than they seemed to be, for their peculiar spectra turned out to consist of lines shifted unbelievably far toward the red. These enormous red shifts meant that quasars are remarkably far away—a billion light-years away and more. They are, in fact, farther off than any of the visible galaxies, and are the most distant objects known. Some quasars are known to be over ten billion light-years away from us.

Some pulsars exist in pairs and revolve about each other. Some quasars exist behind galaxies that lie between us and them. The manner in which pulsars, under such conditions, slow down, and quasars, under such conditions, yield images that seem to split in two or more parts, offers proof that Einstein's general theory of relativity is correct. This is excellent evidence, yet it involves objects not known to exist at the time Einstein evolved his theory.

As for the Universe as a whole, by 1935 we knew that it was expanding, and there was a suggestion that the Universe began many billions of years ago in the form of a giant explosion. This "big bang," however, had no real evidence to support it, and in 1948 a competing theory of "continuous creation" (the slow creation of matter that formed new galaxies between older galaxies that pulled away from each other) was introduced.

It was suggested in 1949 that if the big bang had taken place the fiery energies that existed at the first would spread out and cool down as the Universe expanded. Nowadays, those energies would exist as radio wave radiation coming equally from all parts of the sky—the kind of radio waves that would be expected at a temperature near absolute zero.

In 1964 precisely this sort of radio wave radiation was detected, and now almost all astronomers accept the big bang theory of the beginnings of the Universe.

The details of the big bang are still under discussion, however. The important questions about the big bang deal with the details of events that took place in the tiniest fractions of a second after the actual beginning.

To understand those events, advances had to be made in our understanding of the most basic particles of matter. In the 1970s new theories of these particles were worked out, and these theories were used, in turn, to work out the scenario of what is called "the inflationary Universe," which explains a great many things about the evolution of the galaxies and the general properties of the Universe that had been murky before.

And so from the Earth to the farthest bounds and earliest times of the Universe, our understanding has grown enormously in the last half-century. Subtract all we have learned in that time and our knowledge of astronomy in 1935 would seem small indeed.

Afterword: This essay was written in July 1985 and Voyager 2 *had not yet reached Uranus. That's why I had nothing to say about that planet here.*

10

THE MYTH OF THE MACHINE

To a physicist, a machine is any device that transfers a force from the point where it is applied to another point where it is used and, in the process, changes its intensity or direction.

In this sense it is difficult for a human being to make use of anything that is not part of his body without, in the process, using a machine. A couple of million years ago, when one could scarcely decide whether the most advanced hominids were more human-like than ape-like, pebbles were already being chipped and their sharp edges used to cut or scrape.

And even a chipped pebble is a machine, for the force applied to the blunt edge by the hand is transmitted to the sharp end and, in the process, intensified. The force spread over the large area of the blunt end is concentrated upon the small area of the sharp end. The pressure (force per area) is therefore increased, and without ever increasing the total force, that force is intensified in action. The sharp-edge pebble could, by the greater pressure it exerts, force its way through an object, as a rounded pebble (or a man's hand) could not.

In actual practice, however, few people, other than physicists at their most rigid, would call a chipped pebble a machine. In actual practice, we think of machines as relatively complicated devices, and are more likely to use the name if the device is somewhat removed from direct human guidance and manipulation.

The further a device is removed from human control, the more authentically mechanical it seems, and the whole trend in technology has been to devise machines that are less and less under direct human control and more and more under their own apparent will. A chipped pebble is almost part of the hand it never leaves. But a thrown spear declares a sort of independence the moment it is released.

The clear progression away from direct and immediate control made it possible for human beings, even in primitive times, to slide forward into extrapolation, and to picture devices still less controllable, still more independent than anything of which they had direct experience. Immediately we have a form of fantasy—what some, defining the term more broadly than I would, might even call science fiction.

Man can move on his feet by direct and intimate control; or on horseback, controlling the more powerful animal muscles by rein and heel; or on ship, making use of the invisible power of the wind. Why not progress into further etherealization by way of seven-league boots, flying carpets, self-propelled boats? The power used in these cases was "magic," the tapping of the superhuman and transcendental energies of gods or demons.

Nor did these imaginings concern only the increased physical power of inanimate objects, but even increased mental powers of objects which were still viewed as essentially inanimate. Artificial intelligence is not really a modern concept.

Hephaistos, the Greek god of the forge, is pictured in *The Iliad* as having golden mechanical women, which were as mobile and as intelligent as flesh-and-blood women, and which helped him in his palace.

Why not? After all, if a human smith makes inanimate metal objects of the base metal iron, why should not a god-smith make far more clever inanimate metal objects of the noble metal gold? It is an easy extrapolation, of the sort that comes as second nature to science fiction writers (who, in primitive times, had to be myth-makers, in default of science).

But human artisans, if clever enough, could also make mechanical human beings. Consider Talos, a bronze warrior made by that Thomas Edison of the Greek myths, Daidalos. Talos guarded the shores of Crete, circling the island once each day and keeping off all intruders. The fluid that kept him alive was kept within his body by a plug at his heel. When the Argonauts landed on Crete, Medea used her magic to pull out the plug and Talos lost all his pseudo-animation.

(It is easy to ascribe a symbolic meaning to this myth. Crete, starting in the fourth century, before the Greeks had yet entered Greece, had a navy, the first working navy in human history. The Cretan navy made it possible for the islanders to establish an empire over what became the Greek islands and the Greek mainlanders. The Greek barbarians invading the land were more or less under Cretan dominion to begin with. The bronze-armored warriors carried by the ships guarded the Cretan mainland for two thousand years—and then failed. The plug was pulled, so to speak, when the island of Thera exploded in a vast volcanic eruption in 1500 B.C. and a tsunami destroyed the Cretan civilization— and a Greek civilization took over. Still, the fact that a myth is a sort of vague and distorted recall of something actual does not alter its function of indicating a way of human thinking.)

From the start, then, the machine has faced mankind with a double aspect. As long as it is completely under human control, it is useful and good and makes a better life for people. However, it is the experience of mankind (and was already his experience in quite early times) that technology is a cumulative thing, that machines are invariably improved, and that the improvement is always in the direction of etherealization, always in the direction of less human control and more auto-control—and at an accelerating rate.

As the human control decreases, the machine becomes frightening in exact proportion. Even when the human control is not visibly decreasing, or is decreasing at an excessively low rate, it is a simple task for human imagination to look forward to a time when the machine may go out of our control altogether, so this fear of that can be felt "in advance."

What is the fear?

The simplest and most obvious fear is that of the possible harm that comes from machinery out of control. In fact, any technological advance, however fundamental, has this double aspect of good/harm and, in response, is viewed with a double aspect of love/fear.

Fire warms you, gives you light, cooks your food, smelts your ore—and, out of control, burns and kills. Your knives and spears kill your animal enemies and your human foes and, out of *your* control, is used by your foes to kill you. You can run down the list and build examples indefinitely, and there has never been any human activity which, on getting out of control and doing harm, has not raised this sigh: "Oh, if we had only stuck to the simple and virtuous lives of our ancestors, who were not cursed with this new-fangled misery."

Yet is this fear of piecemeal harm from this or that advance the kind of deep-seated terror so difficult to express that it finds its way into the myths?

I think not. Fear of machinery for the discomfort and occasional harm it brings has (at least until very recently) not moved humanity to more than that occasional sigh. The love of the uses of machinery has always far overbalanced such fears, as we might judge if we consider that at no time in the history of mankind has any culture *voluntarily* given up significant technological advances for the inconvenience or harm of its side-effects. There have been involuntary retreats from technology as a result of warfare, civil strife, epidemics, or natural disasters, but the results of that are precisely what we call "dark ages" and populations suffering from one does their best over the generations to get back on the track and restore the technology they lost.

Mankind has always chosen to counter the evils of technology, not by abandonment of technology, but by additional technology. The smoke of an indoor fire was countered by the chimney. The danger of the spear was countered by the shield. The danger of the mass army was countered by the city wall.

This attitude, despite the steady drizzle of backwardist outcries, has continued to the present. Thus the characteristic technological product of our present life is the automobile. It pollutes the air, assaults our ear-drums, kills fifty thousand Americans a year, and inflicts survivable injuries on hundreds of thousands.

Does anyone seriously expect Americans to give up their murderous little pets voluntarily? Even those who attend rallies to denounce the mechanization of modern life are quite likely to reach those rallies by automobile and would probably think you odd if you objected to that.

The first moment when the magnitude of possible evil was seen by *many* people as uncounterable by *any* conceivable good came with the fission bomb in 1945. Never before had any technological advance set off demands for abandonment by so large a percentage of the population.

In fact, the response to the fission bomb set a new fashion. People were readier to oppose other advances they saw as unacceptably harmful in its side-effects—biological warfare, the SST, certain genetic experiments on microorganisms, breeder reactors, spray cans.

And even so, not one of these items has yet been given up.

But we're on the right track. The fear of the machine is not at the deepest level of the soul if the good the machine does is also accompanied by harm, or if the harm is merely to some people—the few who happen to be on the spot in a vehicular collision, for instance.

The majority, after all, escape, and reap the good of the machine.

No, it is only when the machine threatens all mankind in such a way that each human being begins to feel that he, *himself*, will not escape, that fear overwhelms love.

But since technology has only begun to threaten the human race as a whole in the last thirty years, were we immune to fear before that—or has the human race always been threatened?

After all, is physical destruction by brute energy of a type only now in our fist the only way in which human beings can be destroyed? Might not the machine destroy the essence of humanity, our minds and souls, even while leaving the body intact and secure and comfortable?

It is a common fear, for instance, that television makes people unable to read and pocket computers make them unable to add. Or think of the Spartan king who, on observing a catapult in action, mourned that the device would put an end to human valor.

Certainly such subtle threats to humanity have existed and been seen through all the long ages when man's feeble control over nature made it impossible for him to do himself very much physical harm.

The fear that machinery might make men effete is not yet, in my opinion, the basic and greatest fear. The one (it seems to me) that hits closest to the core is the general fear of irreversible change. Consider—

There are two kinds of change that we can gather from the Universe about us. One is cyclic and benign.

Day both follows and is followed by night. Summer both follows and is followed by winter. Rain both follows and is followed by clear weather, and the net result is, therefore, no change. That may be boring, but it is comfortable and induces a feeling of security.

In fact, so comfortable is the notion that short-term cyclic change promises long-term changelessness that human beings labor to find it everywhere. In human affairs, there is the notion that one generation both follows and is followed by another, that one dynasty both follows and is followed by another, that one empire both follows and is followed by another. It is not a good analogy to the cycles of nature since the repetitions are not exact, but it is good enough to be comforting.

So strongly do human beings want the comfort of cycles that they will seize upon the notion of them when evidence is insufficient—or even when it actually points the other way.

With respect to the Universe, what evidence we have points to a hyperbolic evolution, a Universe that expands forever out of the initial big bang and ends as formless gas and black holes. Yet our emotions drag us, against the evidence, to notions of oscillating, cyclic, repeating universes, in which even the black holes are merely gateways to new big bangs.

But then there is the other change, to be avoided at all costs—the irreversible, malignant change; the one-way change; the permanent change; the change-never-to-return.

What is so fearful about this other kind of change? The fact is that there is one such change that lies so close to ourselves that it distorts the entire Universe for us.

We are, after all, old. We were once young, but we shall never be young again. Irreversible! Our friends are dead, and though they were once alive, they shall never be alive again. Irreversible! The fact is that life ends in death, and that such a change is not cyclic, and we fear that end and know it is useless to fight it.

What is worse is that the Universe doesn't die with us. Callously and immortally it continues onward in its cyclic changes, adding to the injury of death the insult of indifference.

And what is still worse is that other human beings don't die with us. There are younger human beings, born later, who were helpless and dependent on us to start with, but who grow into nemeses who take our place as we age and die. To the injury of death is added the insult of supplantation.

Did I say it is useless to fight this horror of death? Not quite. The uselessness is felt only when we cling to the rational. But there is no law that says we must cling to the rational, and human beings do not, in fact, always do so.

Death can be avoided by simply denying it exists. We can suppose that life on Earth is an illusion, a short testing period prior to entry into some afterlife where all is eternal and there is no question of irreversible change. Or we can suppose that it is only the body that is subject to death, and that there is an immortal component to ourselves not subject to irreversible change, a component that after the death of one body might enter another, in indefinite and cyclic repetitions of life.

These mythic inventions of afterlife and transmigration may make life tolerable for many human beings and enable them to face death with reasonable equanimity—but the fear of death and supplantation is only masked and overlaid; it is not removed.

In fact, the Greek myths involve the successive supplantation of one set of immortals by another—in what seems to be a despairing admission that not even eternal life and superhuman power can remove the danger of irreversible change and the humiliation of being supplanted.

To the Greeks it was disorder ("Chaos") that first ruled the Universe, and it was supplanted by Ouranos (the sky), whose intricate powdering of stars and complexly moving planets symbolized order ("Kosmos").

But Ouranos was castrated by Kronos, his son. Kronos, his brothers, his sisters, and their progeny then ruled the Universe.

Kronos feared that he would be served by his children as he had served his father (a kind of cycle of irreversible changes) and devoured his children as they were born. He was duped by his wife, however, who managed to save her last-born, Zeus, and spirit him away to safety. Zeus grew to adult godhood, rescued his siblings from his father's stomach, warred against Kronos and those who followed him, defeated him, and replaced him as ruler.

(There are supplantation myths among other cultures, too, even in our own—as the one in which Satan tried to supplant God and failed, a myth that reached its greatest literary expression in John Milton's *Paradise Lost.*)

And was Zeus safe? He was attracted to the sea-nymph Thetis and would have married her had he not been informed by the Fates that Thetis was destined to bear a son mightier than his father. That meant it was not safe for Zeus, or for any other god either, to marry her. She was therefore forced (much against her will) to marry Peleus, a mortal, and bear a mortal son, the only child the myths describe her as having. That son was Achilles, who was certainly far mightier than his father (and, like Talos, had only his heel as his weak point through which he might be killed).

Now, then, translate this fear of irreversible change and supplantation into the relationship of man and machine, and what do we have? Surely the *great* fear is not that machinery will harm us—but that it will supplant us. It is not that it will render us weak—but that it will make us obsolete.

The ultimate machine is an intelligent machine. And there is only one basic plot to the "intelligent machine" story—that it is created to serve man, but that it ends by dominating man. It cannot exist without threatening to supplant us, and it must therefore be destroyed or we will be.

There is the danger of the broom of the sorcerer's apprentice, the golem of Rabbi Löw, the monster created by Dr. Frankenstein. As the child born of our body eventually supplants us, so does the machine born of our mind.

Mary Shelley's *Frankenstein,* which appeared in 1818, represents a peak of fear, however, for, as it happened, circumstances conspired to reduce that fear, at least temporarily.

Between the year 1815, which saw the end of a series of general European wars, and 1914, which saw the beginning of another, there was a brief period in which humanity could afford the luxury of optimism concerning its relationship to the machine. The Industrial Revolution seemed suddenly to uplift human power and to bring on dreams of a technological Utopia on Earth in place of the mythic one in heaven. The good of machines seemed to far outbalance the evil, and the response of love to far outbalance the response of fear.

It was in that interval that modern science fiction began—and by modern science fiction I refer to a form of literature that deals with societies differing from our own specifically in the level of science and technology, ones into which we might conceivably pass by imaginable and rational changes in technology. (This differentiates science fiction from fantasy or from "speculative fiction," in which the fictional society cannot be connected with our own by a rational set of changes.)

Modern science fiction, because of the time of its beginning, took on an optimistic note. Man's relationship to the machine was one of use and control. Man's power grew and man's machines were his faithful tools, bringing him wealth and security and carrying him to the farthest reaches of the Universe.

This optimistic note continues to this day, particularly among those writers who were molded in the years before the coming of the fission bomb—notably Robert Heinlein, Arthur C. Clarke, and myself.

Nevertheless, with World War I, disillusionment set in. Science and technology, which once promised an Eden, turned out to be capable of delivering Hell as well. The beautiful airplane that fulfilled the age-old dream of flight could deliver bombs. The chemical techniques that produced anesthetics, dyes, and medicines produced poison gas as well.

And the fear of supplantation rose again. In 1921, not long after the end of World War I, Karl Capek's drama *R.U.R.* appeared, and it was the tale of Frankenstein again, but escalated to the planetary level. Not one monster only was created but millions of "robots" (Capek's word—meaning "workers," mechanical ones, that is). So it was not a single monster turning upon his single creator, but robots turning on humanity, wiping out and supplanting it.

From the beginning of the science fiction magazine, in 1926, to 1959 (a third of a century, or a generation) optimism and pessimism battled each other in science fiction, with optimism—thanks chiefly to the influence of John W. Campbell, Jr.—having the better of it.

Beginning in 1939 I wrote a series of influential robot stories that self-consciously combatted the "Frankenstein complex" and made of the robots the servants, friends, and allies of humanity.

It was pessimism, however, that won in the end, and for two reasons—

First, machinery grew more frightening. The fission bomb threatened physical destruction, of course, but worse still was the rapidly advancing electronic computer. That computer seemed to steal the human soul. Deftly it solved our routine problems, and more and more we found ourselves placing our questions in its hands, and accepting its answers with increasing humility.

All that fission and fusion bombs can do is destroy us; the computer might supplant us.

The second reason is more subtle, for it involved a change in the nature of the science fiction writer.

Until 1959, there were many branches of fiction, with science fiction perhaps the least among them. It brought its writers less in prestige and money than almost any other branch, so that no one wrote science fiction who wasn't so fascinated by it that he was willing to give up any chance at fame and fortune for its sake. Often that fascination stemmed from an absorption in the romance of science, so that science fiction writers would naturally picture man as winning the universe by learning to bend it to his will.

In the 1950s, however, competition with TV gradually killed the magazines that supported fiction, and by the time the 1960s arrived the only form of fiction that was flourishing, and even expanding, was science fiction. Its magazines continued and an incredible paperback boom was initiated. To a lesser extent it invaded movies and television, with its greatest triumphs yet to come.

This meant that in the 1960s and 1970s young writers began to write science fiction not because they wanted to, but because it was there—and because nothing else was there. It meant that many of the new generation of science fiction writers had no knowledge of science, no sympathy for it—and were in fact rather hostile to it. Such writers were far more ready to accept the fear half of the love/fear relationship of man to machine.

As a result, contemporary science fiction, far more often than not, is presenting us with the myth of the child supplanting the parent, Zeus supplanting Kronos, Satan supplanting God, the machine supplanting humanity.

But allow me my own cynical commentary. Remember that although Kronos foresaw the danger of being supplanted, and though he destroyed his children to prevent it—he was supplanted anyway, and rightly so, for Zeus was the better ruler.

So it may be that although we will hate and fight the machines, we will be supplanted anyway, and rightly so, for the intelligent machines to which we will give birth may, better than we, carry on that striving toward the goal of understanding and using the Universe, climbing to heights we ourselves could never aspire to.

Afterword: This essay was written in January 1976. Since then, I have discovered that I was wrong in saying that "at no time in the history of mankind has any culture voluntarily given up significant technological advances for the inconvenience or harm of its side-effects."

It seems that the Japanese in early modern times cut themselves off from the rest of the world and then abandoned the firearms they had adopted in the early seventeenth century. I might have rewritten that paragraph to make myself seem all-wise, but that would be cheating. If I fell into error through ignorance, that might as well be left on display.

However, that bit about Japan was a special case. Japan, effectively isolated, did not have to fear the outside world; the samurai could thus afford to give up firearms. In fact, they had to, if they were to remain supreme. Peasants with guns could too easily kill samurai with those big swords, or even samurai with guns of their own. By restricting everything to big swords, the samurai, properly trained, could lord over the population.

Once Commodore Perry forced Japan to open its gates to the outside world, the Japanese adopted firearms and modern weapons in general in a very great hurry.

As for the ending, in which I imply it might be a good thing for intelligent machines to take over—I later developed a more subtle point of view, which you will see in "Should We Fear the Computer?" (essay #53), which was written in September 1983.

PART II
PRESENT

11

THE PERENNIAL FRINGE

I doubt that any of us really expects to wipe out pseudoscientific beliefs. How can we when those beliefs warm and comfort human beings?

Do you enjoy the thought of dying, or of having someone you love die? Can you blame anyone for convincing himself that there is such a thing as life-everlasting and that he will see all those he loves in a state of perpetual bliss?

Do you feel comfortable with the daily uncertainties of life, with never knowing what the next moment will bring? Can you blame anyone for convincing himself he can forewarn and forearm himself against these uncertainties by seeing the future clearly through the configuration of planetary positions, or the fall of cards, or the pattern of tea-leaves, or the events in dreams?

Inspect every piece of pseudoscience and you will find a security blanket, a thumb to suck, a skirt to hold.

What have we to offer in exchange? Uncertainty! Insecurity!

For those of us who live in a rational world, there is a certain strength in understanding; a glory and comfort in the effort to understand where the understanding does not as yet exist; a beauty even in the most stubborn unknown when it is at least recognized as an *honorable* foe of the thinking mechanism that goes on in three pounds of human brain; one that will gracefully yield to keen observation and subtle analysis, once the observation is keen enough and the analysis subtle enough.

Yet there is an odd paradox in all this that amuses me in a rather sardonic way.

We, the rationalists, would seem to be wedded to uncertainty. We know that the conclusions we come to, based, as they must be, on rational evidence, can never be more than tentative. The coming of new evidence, or of the recognition of a hidden fallacy in the old reasoning, may quite suddenly overthrow a long-held conclusion. Out it must go, however attached to it one may be.

That is because we have *one* certainty, and that rests not with any conclusion, however fundamental it may seem, but in the process whereby such conclusions are reached and, when necessary, changed. It is the scientific process that is certain, the rational view that is sure.

The fringers, however, cling to conclusions with bone-crushing strength. They have no evidence worthy of the name to support those conclusions, and no rational system for forming or changing them. The closest thing they have to a process of reaching conclusions is the acceptance of statements they consider authoritative. Therefore, having come to a belief, particularly a security-building belief, they have no other recourse but to retain it, come what may.

When we change a conclusion it is because we have built a better conclusion in its place, and we do so gladly—or possibly with resignation, if we are emotionally attached to the earlier view.

When the fringers are faced with the prospect of abandoning a belief, they see that they have no way of fashioning a successor and, therefore, have nothing but vacuum to replace it with. Consequently, it is all but impossible for them to abandon that belief. If you try to point out that their belief goes against logic and reason, they refuse to listen, and are quite likely to demand that you be silenced.

Failing any serviceable process of achieving useful conclusions, they turn to others in their perennial search for authoritative statements that alone can make them (temporarily) comfortable.

I am quite commonly asked a question like this: "Dr. Asimov, you are a scientist. Tell me what you think of the transmigration of souls." Or of life after death, or of UFOs, or of astrology—anything you wish. What they want is for me to tell them that scientists have worked out a rationale for the belief and now know, and perhaps have always known, that there was some truth to it.

The temptation is great to say that, as a scientist, I am of the belief that what they are asking about is a crock of unmitigated nonsense—but that is just a matter of supplying them with another kind of authoritative statement and one they won't under any circumstances accept. They will just grow hostile.

Instead, I invariably say, "I'm afraid that I don't know of a single scrap of scientific evidence that supports the notion of transmigration of souls," or whatever variety of fringe they are trying to sell.

This doesn't make them happy, but unless they can supply me with a piece of credible scientific evidence—which they never can—there is nothing more to do. And who knows—my remark might cause a little germ of doubt to grow in their minds, and there is nothing so dangerous to fringe beliefs as a bit of honest doubt.

Perhaps that is why the more "certain" a fringer is, the more angry he seems to get at any expression of an opposing view. The most deliriously certain fringers are, of course, the creationists, who, presumably, get the word straight from God by way of the Bible that creationism is correct. You can't get a more authoritative statement than that, can you?

I occasionally get furious letters from creationists, letters that are filled with opprobrious adjectives and violent accusations. The temptation is great to respond with something like this: "Surely, my friend, you know that you are right and I am wrong, because God has told you so. Surely, you also know that you are going to heaven and I am going to hell, because God has told you that, too. Since I am going to hell, where I will suffer unimaginable torments through all of eternity, isn't it silly for you to call me bad names? How much can your fury add to the infinite punishment that is awaiting me? Or is it that you are just a little bit uncertain and think that God may be lying to you? Or would you feel better applying a little torment of your own (just in case he *is* lying) by burning me at the stake, as you could have done in the good old days when creationists controlled society?"

However, I never send such a letter. I merely grin and tear up his.

But then is there nothing to fight? Do we simply shrug and say that the fringers will always be with us, that we might just as well ignore them and simply go about our business?

No, of course not. There is always the new generation coming up. Every child, every new brain, is a possible field in which rationality can be made to grow. We must therefore present the view of reason, not out of a hope of reconstructing the ruined minds that have rusted shut, which is all but impossible—but to educate and train new and fertile minds.

Furthermore, we must fight any attempt on the part of the fringers and irrationalists to call to their side the force of the state. We cannot be defeated by reason, and the fringers don't know how to use that weapon anyway, but we can be defeated (temporarily, at any rate) by the thumbscrew and the rack, or whatever the modern equivalents are.

That we must fight to the death.

───── 12 ─────

THE CASE AGAINST 'STAR WARS'

I suspect that in future times (assuming there are any) Reagan's administration will be referred to as the "Hollywood Presidency." Certainly, he has all the Hollywood characteristics: an emphasis on image, a winning smile, a voice that oozes geniality, a surface benevolence.

And to top it off he has a tendency to fall for Hollywood solutions. He wants to win the war with something he calls "Strategic Defense Initiative," a harmless phrase he has carefully memorized. Everyone else, with a finer ear for the facts, calls it "Star Wars."

"Star Wars" is a plan whereby devices will be placed in orbit that will continually monitor surrounding space. As soon as an intercontinental ballistic missile with a nuclear warhead, or fifty such missiles, or five hundred, are detected zooming their way from the Soviet Union to the United States, they will one and all be struck and destroyed in their tracks. The United States, totally unharmed, will then face a Soviet Union with its missiles uselessly expended. We will then be in a position to clobber the Evil Empire with our own missiles and wipe it off the face of the Earth. Or, since the United States is good and noble, it will refrain from doing that but will, instead, tell the Soviet Union firmly that unless they change their ways and elect a good Republican Politburo, with perhaps one Democrat in the opposition, we will then clobber them. The Soviet Union will have no choice but to give in; there will be universal peace; and the world will bless Ronald Reagan.

What's wrong with that? Well, we might ask whether "Star Wars" will work.

Naturally, we can't tell in advance, but surely we oughtn't to assume it won't. Forty-five years ago, we determined to devise a nuclear bomb and, by golly, we did it, didn't we? Are our spines made of spaghetti that we are afraid we won't be able to do it again?

Well, the two aren't really comparable.

1. The nuclear bomb was built in deepest secrecy, and there was no real competition. Both Germany and Japan knew about uranium fission, and both had excellent physicists; indeed, uranium fission was actually discovered by German physicists. However, at the time the United States began to work on the nuclear bomb, Germany and Japan were far more involved in war than we

were. Then, as time went on, each month saw them more clobbered than the month before, so that they never really had the ability to put enough effort into such long-distance, uncertain research. To be sure, the Germans came up with rocketry (a far simpler task), and came up with even that too late. The Japanese could do no better than kamikaze pilots.

"Star Wars" research, on the other hand, is being conducted in the open—at a shout, in fact. The Soviet Union may be expected to be in full competition and it is conceivable that it may come up with a working system first. It's not very likely, to be sure, for our technology is far in advance of theirs, and they don't have the "Yankee know-how" which displayed itself so well in the 1940s with Einstein, Bohr, Szilard, Teller, Wigner, and all those other "Yankees."

Just the same, funny things could happen. In the 1950s, when we announced we were going to put objects into orbit, the Soviets said they would, too. We all laughed heartily at that (including me—I laughed like anything). Yet the Soviets got into space first and the United States went into a rather unlovely panic for a time.

What if the Soviets get a "Star Wars" defense into space before we do, and *they* are untouchable while *we* remain vulnerable? Should we labor to try to outlaw the whole thing? To be sure, the Soviets are desperately trying to get us to outlaw it, showing that they have no faith in their own ability to get there first, but what if that's just a double doublecross? Or what if the Soviets, working with desperate energy, surprise themselves as well as surprising us?

Surely, the possibility is worth keeping in mind.

2. The nuclear bomb was a rather simple job. We had to separate uranium-235 from its isotope by known procedures and then we were practically home safe. It was a matter of spending enough money to build large enough "separation systems." Even so, the bomb was just barely made before the war had ended without it. We missed the chance to nuke Stuttgart or Leipzig, and had to make do with Hiroshima and Nagasaki.

"Star Wars," on the other hand, is far more complicated. For one thing, it requires computers, while the nuclear bomb did not. What's more, those computers must be more complex, more rapid, and more fail-proof than any we have now, and the whole thing has to be put into orbit. What's more, we must devise ways of detecting and identifying missiles, telling enemy missiles from harmless objects or from our own missiles, and doing it all without mistakes or omissions.

Can it be done? Let's suppose it can, but then how long will it take? Most people in a position to consider the matter with some expertise seem to think it will take fifty years. And, if there's any lesson we have learned from our recent experience with weapons systems, it usually takes longer to accomplish a task than is estimated beforehand. This means we will have to go along till far into the twenty-first century without a "Star Wars" defense, and a great deal may happen by then. Ought we to put all our efforts and money and emotional intensity into this long range "Star Wars" and leave ourselves less ready to deal with more immediate and perhaps more pressing defense problems?

3. And how much will it cost to put "Star Wars" into action? The nuclear bomb cost us two billion dollars, but it was wartime and no one was inclined to study the bookkeeping of secret war projects. Undoubtedly, if Congress had known about that two billion there would have been an awful howl.

"Star Wars," on the other hand, is not secret, and it's going to cost us something like two billion dollars just to do the initial research (whatever that means) and Congress will have to approve it. It will then take another two billion, or another five billion, to do a little more research, and so on. No one can tell how much it will cost before "Star Wars" is completed and in action, but the guesses I hear run from five hundred billion to a thousand billion dollars. That's half a trillion to a trillion dollars.

Here again, we must remember that from our recent experience with weapons systems, this project will almost certainly take a far greater sum of money to complete the job than is anticipated. Our great corporations, in their patriotic zeal to complete the war projects in which they are engaged, seem to reserve their greatest ingenuity for the task of working out methods for charging the government higher sums and paying the government lower taxes, and the Pentagon seems remarkably willing to let them do so.

We might argue that the huge amount of money required for "Star Wars" is actually a plus, for the Soviet Union will be forced to try to match us in this respect, and its economy is so much weaker than ours that the effort will destroy it financially and lead to a political upheaval that will replace Communist tyranny with something akin to decent, conservative Republicanism. How pleasant it would be if that came about without our having to fire a single shot.

Well, maybe. But meanwhile, we'll be spending billions and tens of billions and hundreds of billions in a "strong economy" that allows us to hold down inflation only through a negative balance of trade that brings in cheap imports and shuts down American industries, and through the ceaseless buiding up of an enormous national debt. The existence of this debt means we will have to pay hundreds of billions of dollars in interest each year to those people, native and foreign, who are rich enough to invest in government bonds, while cutting down drastically on all social services needed by the poor and unfortunate. Reagan's reverse Robin Hood policy of taking from the poor and giving to the rich will be exacerbated by the expense of "Star Wars" and may destroy our own economy before it destroys the Soviet economy.

4. Given the time and the money, will "Star Wars" work?

To be sure, the nuclear bomb worked, but if it hadn't, the only loss would have been the money and effort expended. The nuclear bomb was purely offensive. If, as it had tumbled down through the air toward Hiroshima, it had never gone off, we would have won the war anyway, and possibly only a very little time later and with very few additional lives lost. Those who want to justify the use of the nuclear bomb are constantly speaking of the millions of lives it would have taken to invade the Japanese home islands, but there is at least equal justification in the argument that the Japanese were at the point of surrender, and that (as some maintain) the use of the nuclear bomb was intended as a demonstration, not to the despairing Japanese, but to our too-successful ally, the Soviet Union.

"Star Wars," on the other hand, is a defensive weapon. If it fails, if its protection of the United States against a Soviet attack should prove illusory, then we may be in a desperate situation. Our government, rendered bolder and more likely to take risks by a fancied security, may provoke a nuclear war that will destroy us as well as the Soviet Union, whereas without "Star Wars" we might have followed a more cautious foreign policy that might have achieved American aims without nuclear war.

Nor might "Star Wars" prove a failure only because of inherent short-comings. It might do everything we dream of and yet turn out to be insufficient even so. Suppose the Soviet Union despairs of building a "Star Wars" of its own, and concentrates instead on building some offensive weapon that can pierce the "Star Wars" defense. Such an offensive weapon may be quite simple and may cost very little and the Soviets will have, very likely, fifty years to dream it up and put it in action. "Star Wars" will then become another Maginot Line for future historians to shake their heads over (assuming there will be future historians).

But never mind, let us be very imaginative and let us suppose that "Star Wars" is put in place—that it didn't take too much time and didn't cost too much money and that it works perfectly. Let us suppose also that the Soviets have failed to match us and have worked out no counterweapon. Where would we stand?

It would seem that we would be in a wonderful position. We would be invulnerable and we could destroy the Soviet Union at will without fear of reprisal. What could be better?

In fact, during the recent presidential campaign, Reagan seemed to feel so euphoric over such a situation that he said, on two different occasions, that we would offer the Soviets the "Star Wars" technology, once we had set up our defense. "Why not?" he asked. "Why not?" and Mondale had no effective or expedient answer.

I have no problems with expediency, so here's why not!

If the Soviet Union and we *both* had an effective "Star Wars" technology, then each would be invulnerable to the other's long range, space-penetrating missiles. Each, however, would have intermediate range missiles that go through the atmosphere and that might work despite "Star Wars." Each would also have planes, tanks, submarines, and all sorts of non-nuclear war technology. And the Soviet Union might still be the Evil Empire.

In that case, a war might be fought in which only the long-range missiles would not be used.

But it is the long-range missiles that are the chief weapon in our arsenal and the best way of wiping out the Soviet Union in half an hour. With them gone, we might face a long, partly conventional war—and we don't really enjoy such things, as our experience in Korea and Vietnam have amply demonstrated.

So we can't, and *won't*, give the Soviets the "Star Wars" technology, if we have it and they don't.

Why, then, did Reagan suggest we would? The unkind explanation would be to suggest that he hadn't thought it through, or wasn't capable of thinking it through. A kinder explanation would be to suggest he was only playing his Hollywood "good guy" role, and has no real intention of making such a gift, even if he were to live long enough to see the project completed and were still in a position of power.

So forget that and let's go back to the imagined situation in which we have "Star Wars" while the Soviets don't. What's to stop us, under such circumstances, from bringing about world peace and eternal happiness by threatening to destroy the Soviet Union if they don't behave themselves?

Well, there is such a thing as a "nuclear winter."

Every once in a while a volcano throws dust into the stratosphere, enough dust to reflect a lot of sunlight and keep it from reaching the Earth. That results in a kind of chill on our planet. In 1815 a volcano exploded in the East Indies, and in 1816 New England had snow at least once in every month, including July and August. The year was called "The Year Without a Summer" and "Eighteen Hundred and Froze to Death."

Studies of volcanic action in the nearly two centuries since has showed how dust and smoke in the upper atmosphere might affect world climate. Close-up studies in the 1970s of planetary dust storms on Mars have added to our knowledge.

There is an increasingly strong opinion among many scientists that even a moderate nuclear exchange will throw up a great deal of dust and ash into the stratosphere (that's what the familiar "mushroom cloud" consists of). In addition, it will cause fire storms as forests and cities burn, and that will send vast quantities of ash and soot into the stratosphere. The net result will be that sunlight will be cut off, the Earth will go through a period of darkness and freezing temperature, plant life will die, animals (and human beings) will starve. That is the nuclear winter.

Conservatives tend to maintain that the notion of the nuclear winter is based on a variety of unlikely assumptions, but how are we going to test the matter? Have a nuclear war and see if we kill the human race, or only half of it?

As a matter of fact, even the Pentagon has now been forced to admit that a nuclear winter seems to be a likely possibility in case of nuclear war.

Well, then, how are we going to threaten the Soviet Union from behind our "Star Wars" defense? Will we say, "We'll blast you into oblivion unless you surrender unconditionally right now?"

What if they answer, "Go ahead. You too will be destroyed"?

It may be that this is what bothers conservatives about a nuclear winter. It makes an overwhelming first strike impossible and destroys what is, perhaps, a favorite dream of theirs with regard to the Soviet Union.

It seems, then, that "Star Wars" will not accomplish anything for us, even if we have the time to build it, even if we can afford it, and even if it works as we think it ought to—none of which is very likely. "Star Wars" is, in fact, only Hollywood science fiction, and, like almost all Hollywood science fiction, it is bad science fiction.

What, then, ought to be done?

It seems to me that the world must give up the idea of war altogether. Modern war is world suicide and once this is thoroughly understood, there will perhaps be a determined search for ways of limiting weapons of all kinds; better yet, for ways of reducing weapons of all kinds; and, best of all, for abolishing weapons of all kinds.

We need disarmament; that's what we need. We need international cooperation; that's what we need. We need a new world realization that there is only one human species, *Homo sapiens,* and that we must all stand together against the problems that face us all; that's what we need.

And if we can't achieve disarmament, cooperation, and a feeling of world unity, then that is too bad, for with leaders possessing the kind of mentality typified by the "Star Wars" fantasy, we are not going to survive for much longer.

Afterword: This essay, when it first appeared, elicited angry letters, very much like those I received fifteen or more years ago when I said the Vietnam War was a stupid mistake.

SHORT TERM; LONG TERM

The nuclear disaster at Chernobyl in the USSR gave the United States a terrific opportunity to play for short-term advantage—at the expense, however, of long-term trouble.

The Soviet Union tried hard to stonewall and to play down the incident. This, for a number of obvious reasons. First, Russia, under whatever form of government, has a long tradition of stolidity in the face of disaster. The Soviet government, in particular, emphasizes its efficient and scientific character. Its officials would like us to believe that disasters do not happen there, or are speedily corrected when they do happen.

So the government announced the incident only when heightened radioactive levels in Sweden made it impossible to hide. It stated that two people had died, eighteen were in serious condition, and a few dozen others were affected less seriously. The situation, government officials claimed, was almost immediately brought under control.

Clearly, they were trying to put the best face on the situation, and the United States had no intention of allowing them to do so. It was in the short-term interest of the United States to frustrate Soviet aims.

The American media and government from the start insisted that the disaster was much worse than the Soviets admitted, and that, far from being under control, it was continuing unabated. American television and newspapers assiduously but irresponsibly spread unconfirmed reports that thousands had been killed, a statement that was quickly converted to hundreds, and then to comments that the death toll might never be known. Our journalists claimed that even if the Soviet statements on casualties were right, the disaster was still much worse than that government admitted.

The fire was reported to be burning out of control. On the evening news programs I watched experts reported that graphite fires were almost impossible to extinguish. One predicted that it would burn two weeks, another four weeks, and stil another that it might burn for six weeks. The very next day, when satellite photos did not unmistakably show the fire—even the earlier photographs were open to alternate interpretations, according to non-American experts—it was said to be smoldering. Soviet photographs that indicated that no fire existed were dismissed as obvious fakes.

The Reagan Administration's purpose to this is threefold, it seems to me.

First, it is meant to show that under the Soviet system, governmental intelligence, efficiency, and safety precautions are so low that disasters such as this could not fail to happen. Second, that the Soviets are so callous to human life that they minimized the disaster even though it meant that their own people were put at unnecessary risk. Third, that the Soviets were, in any case, pathological liars whose word could never be trusted.

Similarly, the rise in radioactivity in neighboring countries has been carefully reported. The precautionary activities in nearby nations, especially those in the Soviet bloc, are detailed. There is the use of potassium iodide doses for children, the advice to avoid water or milk or fresh vegetables, and the recommendation that children stay indoors and cattle be kept in barns.

All this, one can easily see, would tend to rouse European anger against the Soviet Union for the way it put its neighbors at risk by avoiding a full statement of what was going on. And it might be hoped that it would particularly anger such nations as Poland and Rumania, which may thus be encouraged to rebel against Soviet domination.

All this might be viewed by some as clever psychological warfare ploys. And yet what they may achieve is mere short-term victory, which the Soviet Union can survive if it simply holds on (just as it survived the shooting down of the Korean airliner). The long-term effects, however, may not be at all to American liking.

For one thing, the Chernobyl disaster may well lead to a distrust of nuclear power stations everywhere—and particularly in the United States. Immediately after the Three Mile Island disaster, two-fifths of Americans were polled as being opposed to the building of nuclear power stations, while the most recent poll, after Chernobyl, rasied the figure to three-fifths (despite the fact that those bemoaning the Soviet disaster are insistent on the fact that a Chernobyl-type event can't happen here).

The long-term effect may be, then, that nuclear power projects, crippled almost to the point of death by Three Mile Island, may be utterly dead in the water in the United States, while they continue to be developed in the Soviet Union, with (one can presume) more stringent safeguards.

Secondly, the news media in the United States have gone out of their way to cultivate a mood of panic in the world by emphasizing the broad sweep of the radioactive clouds, with maps showing them being carried westward over Europe and eastward over the Ukraine. The references to contaminated water sources, rising cancer rates, and ruined farm land all feed the panic. The hasty evacuation of some Americans studying in the Soviet Union somehow brings it home in such a way that Americans are beginning to worry about the effects of the radioactive cloud when it reaches the United States, even though government spokesmen are carefully stressing the fact that the cloud will be utterly harmless by the time it crosses our border.

This mood of panic may produce this result—

People may think that if the meltdown of a single nuclear reactor can produce so much harm hundreds and thousands of miles from the place of the accident, and if a disaster of this sort is so horribly dangerous even when every effort is made to contain it—then what happens if a nuclear bomb explodes? Or not just one, but hundreds?

In other words, what if there is a nuclear war under the most "favorable" conditions? What if the United States strikes first, in such force and with such accuracy, that the capacity of the Soviet Union to retaliate is utterly destroyed, that the Soviet Union loses the war in the first half-hour, and that its remnant must humbly sue for peace, never capable of troubling the world again?

Yet if *one* out-of-control reactor can so endanger Europe, what will the full-scale explosion of hundreds of nuclear bombs do? How will the radioactive cloud, hundreds, thousands, many thousands of times as intense as the Chernobyl cloud, affect other nations—even the United States? Will not the same blow that destroys the Soviet Union also destroy Western Europe a few days later and virtually the rest of the world besides?

As soon as this view penetrates, the anxiety of Europe to avoid a nuclear war at any price may double and redouble endlessly. The pressure to achieve weapons control and to negotiate international understanding may become irresistible, and the Reagan foreign policy may become untenable.

It may then occur to some American officials that the quick make-them-squirm reaction to Chernobyl was not perhaps the wisest.

Afterword: And, as a matter of fact, five months after this essay was written, Reagan was so carried away by popular opinion in favor of a solution to the nuclear arms race that he hastily agreed to see Gorbachev in Iceland and almost agreed to virtually complete nuclear disarmament. Only his mad passion for "Star Wars" kept him from doing that. American conservatives are still in a state of shock over that narrow escape.

——— 14 ———
THE USEFUL IVORY TOWER

The American taxpayer has paid a billion dollars to have two probes land on the surface of Mars and, in exchange, has seen pictures of that arid world. The same taxpayer has paid perhaps forty billion dollars over a period of ten years for the Apollo program and has received Moon rocks in return.

Why? So that some scientists in their ivory towers can be happy at public expense? Could not the money have been used to better purpose in a thousand other ways—for bettering the schools, for instance?

This all-too-common indignation towards space exploration is wrong in two ways.

In the first place, the indignation is highly selective.

At its peak, space exploration cost each American twenty dollars per year—about as much as we spend on cosmetics for teen-age girls.

At the same time, the American public spent eighty dollars per person per year on tobacco and one hundred dollars per person per year on alcoholic beverages.

To be sure, these other expenses, which the American public bears uncomplainingly and without worrying about the better uses to which the money can be put, bring personal pleasure, in addition to disease and death—but does the space program bring us nothing?

It brings us knowledge, and knowledge can be useful. In fact, it cannot fail to be useful.

All history shows us that knowledge, however useless it appears, can have its applications, sometimes in surprising and unlooked-for ways. All history also shows us that hardly anyone believes this and that all of history's experience will not prevent the person who is satisfied with his ignorance from deriding and denouncing the desire for knowledge in others.

The first case of this that has come down to us is some twenty-three centuries old. About 370 B.C., a student is supposed to have asked the Greek philosopher Plato of what use were the elaborate and abstract mathematical theorems he was being taught. Plato at once ordered the student to be given a small coin so that he might not think he had gained knowledge for nothing—and then dismissed him from the school.

Of course, Plato was proud of the fact that his mathematics were theoretical and had no obvious uses in everyday life, since there was a sign of mental aristocracy in this.

Plato's school was called "the Academy," and the term "academic" is still used for knowledge that seems to have no practical use.

Plato and his student were alike wrong, however. Mathematics, sometimes astonishingly esoteric mathematics, can have its uses. Time and time again in these modern centuries mathematics has proved to be the backbone of technological advance.

Can one tell in advance what knowledge will be useful some day and what knowledge will not? If we could, we might concentrate entirely on the useful direction—or at least spend public money only on the useful direction and leave the scientist to follow fascinating uselessness out of his own pocket.

The trouble is, we can't.

In the 1670s, a Dutch draper named Anton van Leeuwenhoek indulged in his hobby of painstakingly grinding perfect little lenses, which he used to magnify everyday objects. He peered through those lenses constantly and drew pictures of what he saw. He carefully studied ditch water through his magnifying lenses, for instance.

Fortunately, it wasn't an expensive hobby, because he surely would have had trouble getting a government grant for it if his government had grants in those days. The early Dutch equivalent of watch-dog Senators would have had a lot of fun over van Leeuwenhoek's penchant for staring at ditch water.

Yet in 1677 van Leeuwenhoek discovered small living creatures, too small to see without magnification, swimming about in the ditch water. He uncovered a whole unsuspected world of micro-life. And in 1683 he caught his first glimpse of even smaller objects, which later came to be called bacteria.

Purely academic knowledge! How could these findings ever be useful!

Except that two hundred years later, in the 1860s, the French chemist Louis Pasteur showed that these micro-organisms were the cause of infectious disease. It was with that "germ theory" of disease that physicians became, *for the first time,* true healers and began to cure sickness. The death rate dropped and in the last century, over much of the world, the life expectancy has increased from thirty-five to seventy.

The average human being in many nations now lives twice as long as he or she did in Abraham Lincoln's time, because a Dutch draper once peered through tiny lenses.

How much is that worth? Who would not now, in hindsight, be willing to support the Dutch draper out of the public purse, to any extent?

Experience came to show scientists that however attractively academic a scientific discovery might be, the real and grubby world of usefulness was bound to break in. Plato's pride could no longer be supported.

There is a story about the English scientist, Michael Faraday, that illustrates this.

Faraday was, in his time, an enormously popular lecturer as well as a physicist and chemist of the first rank. In one of his lectures in the 1840s, he illustrated the peculiar behavior of a spiral coil of wire which was connected to a galvanometer that would record the presence of an electric current.

Left to itself there was, of course, no current in the wire. When a magnet was thrust into the hollow center of the spiral coil, however, the needle of the galvanometer moved to one side of the scale. When the magnet was withdrawn from the coil, the needle flipped in the other direction, showing that the current was now flowing in the opposite direction. When the magnet was held motionless in any position within the coil, the needle was also motionless, showing that there was no current flowing.

At the conclusion of the lecture, one member of the audience approached Faraday and said, "Sir, the behavior of the magnet and the coil of wire was interesting, but of what possible use can it be?"

And Faraday answered politely, "Sir, of what use is a newborn baby?"

The tale may be apocryphal, for the same final statement is attributed to Benjamin Franklin at the first demonstration of a balloon in Paris in 1783. There is also a second version of the story, in which a leading figure of the British government asks Faraday the question, and Faraday replies:

"Sir, in twenty years you will be taxing it."

Apocryphal or not, however, the story makes sense. It was this phenomenon of coil of wire and moving magnet which Faraday made use of to develop the electric generator. For the first time, it became possible to produce electricity cheaply and in quantity and that, in turn, made it possible to build the electrified technology that surrounds us today and without which life, in the modern sense, is inconceivable. Faraday's demonstration was a newborn baby that grew into a giant and generated not only electricity but enormous quantities of tax money.

Nor is it only the layman who can't judge what is useful and what is not. Even the shrewdest of men can fail.

Perhaps the most ingenious man in history was Thomas Alva Edison, and there never was a man so keen in judging the useful as he. In fact, he was the living embodiment of the "of-what-use-is-it?" school and viewed the academic with a mixture of amusement and contempt.

He learned a lesson in uselessness in 1868 with his first patented invention. It was a device to record votes mechanically. By using it, congressmen could press buttons and all their votes would be instantly recorded and totalled. There was no question but that the invention worked; it remained only to sell it. A congressman whom Edison consulted, however, told him that there wasn't a chance of the invention's being accepted, no matter how unfailingly it might work.

It seemed, you see, that a slow vote was sometimes a political necessity and Congress would not give up that option.

Edison did not make that mistake again. He decided never to invent anything unless he was sure it would be needed and wanted, and not be misguided into error merely because an invention worked.

He stuck to that. Before he died, he had obtained nearly thirteen hundred patents—three hundred of them over a four-year stretch or, on the average, one every five days. Always, he was guided by his notion of the useful and the practical.

On October 21, 1879, he produced the first practical electric light, perhaps the most astonishing of all his inventions. In succeeding years, Edison labored to improve the electric light and find ways of making the glowing filament last longer before breaking. As was usual with him, he tried everything he could think of. One of his hit-and-miss efforts was to seal a metal wire into the evacuated electric light bulb. The metal wire was near the filament but did not touch it so that the two were separated by a small gap of vacuum.

Edison then turned on the electric current to see if the presence of a metal wire would somehow preserve the life of the glowing filament. It didn't and Edison abandoned the approach. However, he could not help noticing that an electric current seemed to flow from the filament, across the vacuum gap, to the wire.

Nothing in Edison's vast practical knowledge of electricity explained that, and all Edison could do was to observe the phenomenon, write it up in his note-books and, in 1884 (being Edison), patent it. The phenomenon was called the "Edison effect" and it was Edison's only discovery in pure science.

Edison could see no use for it. He therefore let the matter go, while he continued the chase after what he considered the useful and practical.

In the 1880s and 1890s, however, scientists who pursued "useless" knowledge for its own sake discovered that subatomic particles (eventually called "electrons") existed, and that the electric current was accompanied by a flow of electrons. The Edison effect was the result of the ability of electrons, under certain conditions, to travel unimpeded through a vacuum.

The Edison effect, improved by further refinements, was used to amplify and control with great delicacy the flow of electrons, and thus was developed the "radio tube." It was the radio tube that first made all our modern electronic devices possible, including television. The Edison effect, which the practical Edison shrugged off as interesting but useless, turned out to have more astonishing and near-miraculous results than any of his practical devices.

In fact, consider the reverse. Suppose scientists, in a Platonic mood, deliberately *wanted* to find a useless branch of research so that they could retire to an ivory tower which nothing would disturb.

Between 1900 and 1930, for instance, theoretical physics underwent a revolution. The development of quantum mechanics led to a new and more subtle understanding of the behavior of the inner components of atoms.

None of it seemed to have the slightest use to humanity, and the scientists involved, a brilliant group of young men, looked back on this time in later life as a happy time of abstraction and impracticality, as a Garden of Eden out of which they had been evicted.

For they *were* evicted. Reality obtruded. Out of their abstract work came the nuclear bomb.

We can make a general rule that *any* advance of knowledge has in it the potential for use by human beings for some purpose or other, for some greater control of the universe, for some betterment of the human condition.

Our methods for reaching the Moon date back to the academic research of an English scientist named Isaac Newton, who in the 1680s worked out the laws of motion.

And now that we have reached the Moon, human beings who are now living may see it become a complex mining base in their own lifetime, one that will produce the raw materials out of which solar power stations may be built in space that will convert solar radiation into a useful and clean energy supply for all the Earth.

Ought we then to support the search for knowledge, however academic it may seem?

Why, of course! Even if only one percent proves useful, that one percent will be enough to justify the entire investment.

But wait! Will knowledge be useful? Nothing else?

The academic work on nuclear physics produced the nuclear bomb—is that useful? Does it not threaten the world with destruction? Is not all our technology a way of poisoning the Earth with pollution? Do not our machines simply point the road to dehumanization?

It depends on what use we make of all this. The nuclear bomb can be produced, but it need not be used for destruction. The same knowledge that gave rise to the nuclear bomb also gave rise to the radioisotopes that have led to important advances in every branch of science.

Technology, unwisely used, can pollute, dehumanize, destroy. Wisely used, it can minimize and reverse pollution, prevent the dehumanization of unremitting labor, bar the destruction of disease and famine.

The boundary between the useful and the dangerous is not very easy to see, remember. In fact, just as practical men throughout history have found it difficult to tell the useful from the useless, so our idealists today may find it difficult to tell the useful from the harmful.

What, for instance, is the most dangerous discovery in human history? Is it the nuclear bomb? No, that may never be used. Is it genetic engineering? No, its dangers may be greatly exaggerated. Is it the pollution-laden internal combustion engine? No, that may be cleaned up.

But what about the development of the germ theory of disease by Louis Pasteur, which I mentioned earlier as having led to medical advances that have doubled our life span?

No one thinks of that discovery as having been dangerous. No one objects to living longer. The idealists who march in protest against nuclear power stations, who demand that pesticides be banned, who insist that this or that chemical be removed from the environment, or that this or that dam not be constructed—never carry banners reading "Down with cancer research" or "Reinstate epidemics." On the contrary, in denouncing the expense of the space program they frequently suggest that the money could better be spent on medical research.

But it is precisely the advances in medicine that have lowered the death rate precipitously in the last century and have greatly increased the population growth rate. Because of this, we now have a world population of four billion, twice the number that existed on Earth when I was born, and a continuing increase of eighty million a year.

If this goes on (and there seems no way of stopping it quickly), there will be mass starvation before the century is over and civilization may not make it into the twenty-first century.

The nuclear bombs may not go off, pollution may be cleaned up, and new energy sources may be in the offing, but what can save us from the population increase? So is not then the discovery of the germ theory of disease the most dangerous single discovery ever made?

But why was it dangerous? Because it was wrong to lengthen human life and ease the scourge of sickness?

Surely not! But it was wrong to use the discovery unwisely.

If the death rate was decreased, the birth rate should have been decreased, too. Unfortunately there are no traditions, either social or religious, against lowering the death rate, while there are strong traditions against lowering the birth rate. Hundreds of millions who eagerly accept the former stand adamant against the latter—and it is this insanity, and not medicine, that is destroying us.

It would seem there are two rules:

1. Support your local ivory tower, for *all* knowledge may be useful, and some will certainly be useful; it is almost impossible to tell in advance what the uses may be or from which direction they will come.

2. Be very cautious in making *any* use of knowledge, for all uses can have their harmful side-effects, and some certainly will; it is almost impossible to tell in advance what the dangers may be or from which direction they will come.

Does Rule 2 cancel out Rule 1 and make the whole game useless?

Not at all. Rule 2 *modifies* Rule 1, and makes the whole game merely risky.

But *all* life is risky, and many people think it wouldn't be interesting otherwise.

15

DO IT FIRST!

Kennesaw, Georgia, passed an ordinance in 1982 to the effect that the head of every household must invest in a gun and ammunition, and keep the stuff around the house. The idea is to scare off criminals.

Of course, if this notion is to work, the ammunition has to be put into the guns and the head of the household has to go through a course in target practice and quick draw. Then he's set.

Clearly, this is in the great American tradition of do-it-yourself—and do it first. Why wait for a criminal to come into the house and shoot someone in your family? If you lose your temper with your wife, or if she loses hers with you, either one of you need only grab up the gun and save the criminal a heap of trouble. Or, if you're careful to leave the gun where your kid can find it, he can shoot a good friend of his; or, with just a little luck, he might even shoot daddy or mummy.

This is an inspiring achievement, which we can credit entirely to American grit and know-how. What's more, it points the way to the solution of a great many other social ills, ills that would otherwise seem totally intractable.

How about littering, for instance? It is always disgusting to see the streets filled with shards of soda pop bottles, fragments of old newspapers, bits of assorted garbage, fecal detritus left behind by wandering canines, and so on.

However, it's only do-gooders and pinko liberals who think they can change human nature. People have always littered; it's the way they are. So beat them to the punch. After all, what really bothers you is that it's not *your* litter and that it therefore lacks that esthetic quality that your litter would naturally have.

Why not pass an ordinance, then, that every week the inhabitants of each city block appoint a committee to go out and litter the sidewalk and roads with selected scraps. If you live in a posh neighborhood, make sure those decayed newspaper scraps are from the *New York Times;* choose broken bottles that once held champagne; have those dog droppings come from pedigreed poodles. Think of the tone it would give the neighborhood to have dented cans of caviar and *pâté de foie gras* spread tastefully over the lawn.

And it would keep criminal litterers away. Their hearts would break, for they would see quite clearly that they could never match the kind of work a neighborhood *that cares* can do.

Very much the same thing could be said about graffiti. Does it really bother you that the subways and public buildings are covered with vague scrawls and misspelled obscenities and drawings of sexual perversions so poorly designed that one can scarcely recognize which particular one is intended? Of course! It's all so inartistic!

Do it first! Let that slogan ring out everywhere. Let's have public-spirited citizens with a sense of color and design turn out in response to city ordinances. Let us march in parade, singing patriotic songs, and swinging our spray cans high. Let us use color tastefully, and let us be literate, grammatical, and orthographic. How impressive it will be to have City Hall covered with a large graffito in early English lettering that reads: "Mayor Koch is nonhirsute!" How much preferable to the usual illiterate scrawls.

Individual householders, at such time as they become tired of playing with their guns, can go out and graffitize their own houses. Unquestionably, the ignorant and unskillful louts to whom we now entrust the task of placing graffiti here and there will expire in shame and frustration, but why should we care what happens to them?

Drunken driving! Now there's a cause that gets all the do-gooders hot and bothered. How can we stop it? Remember prohibition? It can't be done just by passing laws against it. No, sir. Let's pass a city ordinance requiring everybody to have a couple of good slugs of whiskey, gin, or rum before getting behind the wheel.

In the first place, the papers are forever telling us about a drunk driver who plows into a crowd, or into another car, and kills six people. The drunk driver is never hurt. Oh, maybe he gets a little bump on the forehead, but that's about all. Well, if everyone's drunk, everyone gets those little bumps at most, and no one's ever hurt.

Air pollution. Just fix it that every house has a smoky furnace and every car burns oil. Have them all examined every six months and impose a stiff fine on anyone who owns anything that burns clean. That will remove the chief complaint: that your air is dirty while someone else's is clean. Once everyone has dirty air, there'll be no ground for social jealousy and none of us will really mind.

And let's pass a law that everyone has to use fluorocarbon spray cans. Let's get rid of that ozone layer once and for all, instead of having it up there blackmailing us with its constant threat of disappearing. Who needs it?

As for nuclear weapons, that's the simplest problem of all. Everyone wants to get rid of them, and what is the surest way of getting rid of them?—That's right; you explode them. Let's have the United Nations pass a resolution that every nation with a nuclear arsenal explode every one of them immediately. That would not only get rid of those bombs, it would solve all our other problems, too. Without exception.

Let's all give a rising vote of thanks to the people of Kennesaw, Georgia, for showing the way.

Afterword: This was written at the request of an editor who wanted irony. So I gave him irony—as bitter as I could make it. He accepted it and paid for it. He was then replaced as editor and the new editor did not have the same taste for irony, so the item was never published. Following my usual policy, I therefore include it here. I waste nothing.

—— 16 ——

POPULARIZING SCIENCE

In 1686 a middle-class Frenchman, Bernard le Bovier de Fontenelle, was a failure. He had entered his father's profession of the law and it hadn't worked out. He decided to be a writer and had tried his hand at poetry, opera, and drama, and had done poorly in each case.

But now, at the age of twenty-nine, he published a book entitled *Conversations on the Plurality of Worlds,* and with that he hit the jackpot. It was an introduction, for the interested and intelligent layman, to the new astronomy of the telescope; a careful consideration of each of the planets from Mercury to Saturn, with speculations as to the kind of life that might be found upon them. The book was devoured by the upper and middle classes (who alone were literate) and went through many editions. Fontenelle was perhaps the first person to make a reputation in science on the basis of his popular science writing alone.

He was elected to the French Academy in 1691 and became perpetual secretary of the French Academy of Sciences in 1697. He wrote annual summaries of its activities and composed obituaries for famous scientists as they died. He loved society and found himself welcome everywhere. He lived in good health and kept his faculties into advanced old age, dying one month before his hundredth birthday.

There have been science popularizers ever since, including some important scientists—from Davy and Faraday, through Tyndall, Jeans, and Eddington, to Sagan and Gould among our contemporaries.

But why bother? Few popularizers become wealthy, and there is certainly no great demand for it by the general public. Fontenelle wrote in the "Age of Reason," and it was then chic for the aristocracy to allow themselves to be instructed in the new science. But now?

Well, science popularization today is essential, far more so than at any time in the past. Where, in Fontenelle's time, science popularization was entertainment for the otherwise bored, it is now life and death for the world, for four reasons.

1. Science, together with its practical sister, technology, have taken over the world, both for better and for worse. Advances are coming at an ever-accelerating pace and the world is changing with breath-taking speed. If we consider only the last forty years, we have television, the jet plane, the rocket, the transistor, the computer, each of which has completely altered the world we live

in. Lying immediately ahead is the robotization of industry, the computerization of the home, and the expansion into space.

All these things, used wisely, can bring about a much happier world than the one we live in now or have ever lived in—a world more secure, more comfortable, more amusing. All these things, used unwisely, can destroy civilization altogether.

We must not view science and technology as either an inevitable savior or an inevitable destroyer. It can be either, and the choice is ours. A general public, utterly ignorant of science, led by rulers scarcely better informed, cannot be expected to make intelligent choices in this matter. The alternatives of salvation and destruction depend, in this case, upon the blind gropings of ignorance.

This is not to say that we must build a world of scientists. That would be beyond the abilities of popularizers. But at least let the public make up an intelligent and informed audience. Football games are watched by millions who cannot play the game themselves, or even direct one successfully—but who can at least understand enough to applaud and to groan at the proper places.

2. Science is no longer the province of the well-to-do amateur who can finance his explorations out of his own pocket. The days of Henry Cavendish, Lord Rosse, and Percival Lowell are gone. Science is expensive and is growing more so. Continued advance is, to a large extent, dependent upon the support of large industries and of governments. In either case, it comes out of the public purse in the last analysis; and whether that purse will open or not depends upon public willingness to endure the expense.

A public that does not understand how science works can all too easily fall prey to those ignoramuses who make fun of what they do not understand, or to the sloganeers who proclaim scientists to be the tools of the military. The difference to the public between a scientist and a magician is the difference between understanding and not understanding, which is also the difference between respect and admiration on the one side, and hate and fear on the other.

Without an informed public, scientists will not only be no longer supported financially, they will be actively persecuted. Who, then, is to supply the understanding if not the science popularizers?

3. Scientists do not form a closed caste. They do not inherit their calling. New scientists must be recruited from outside, especially since the number of scientists, from decade to decade, is increasing at a more rapid rate than the number of people generally is. How then is the recruitment to take place?

Some youngsters are drawn to science willy-nilly by an inner compulsion, and cannot be kept out of it, but surely the numbers of these scientists-despite-themselves simply will not be great enough. There must be those who are attracted if some stimulus is applied, but perhaps not otherwise. An effective piece of science popularization is surely one way of rousing a spark of interest in a youngster, a spark that may eventually burst into flame.

I daresay there is not a science popularizer in the world who has not received a gratifying number of letters from young readers who explain that they are majoring in physics (chemistry, biology, mathematics, geology, astronomy) and that the original push came from a book that the popularizer had written.

4. There have always been, throughout history, class differences in society. Even when a particular society labors to abolish those differences and to proclaim all individuals equal before the law, natural differences will arise. Some differences are physical. A keen eye, razor-sharp reflexes, a marvellous voice, or a talent for writing or for acting, simply can't be handed out to everyone.

And some differences are imposed by circumstance. Of two people of equal potential, one may receive a better education than the other, or one may gain more practical experience than the other. And although the race may not always be to the swift, it is wise to place your bets on the swift.

As the world progresses and becomes more permeated with science and technology, as it becomes more "high-tech," the type of education that will become ever more necessary for individual advance will be that in science. We are, in the future, running the risk of creating a world of science-educated "haves," whose children will automatically have the opportunity of such education; and science-uneducated "have-nots," whose children are not likely to have the opportunity. If one believes, as I do, that a strongly stratified society has the seeds of instability and destruction in it, then we should labor toward making science education as broadly based as possible, so that as few people as possible are fated from birth to be part of an uneducated underclass.

I would like to conclude now by saying that it was a consideration of all these compelling reasons that led me some quarter-century ago to abandon my academic duties to become a full-time science writer, but I cannot. The truth is that I enjoy writing more than I enjoy anything else, and I would have made the switch if there were no other reason for it at all.

THE PACE OF RESEARCH

The question posed me is this: "Should we slow our scientific research efforts for ethical and/or economic reasons?"

The answer to that is as firm a negative as it would have been if the question were "Should we slow our breathing rate during exercise for oxygen conservation reasons?"

We can't slow our breathing rate during exercise, because the needs of exercise compel us, willy-nilly, to increase our breathing rate. We can't slow our scientific research efforts for ethical and/or economic reasons, because the requirements of ethics and economics constantly demand solutions to problems, solutions that can only come from scientific research.

What ethical and/or economic considerations are we thinking of?

How do we feed the world's hungry, clothe the naked, house the homeless, heal the sick? How do we educate our children, protect our aged, preserve our environment, abolish tyranny, and secure our freedom?

Well, how can we do any of these things unless we learn how to handle the resources of Earth more efficiently and wisely than we now do? Unless we learn to conserve those resources and tap new sources of energy? Unless we learn how more efficiently to organize society in such ways as to control our dangerous population growth and defuse our undesirable and meaningless social injustices and group antagonisms?

And how can we do any of these things without continuing scientific research?

To be sure, scientific research leads to technological advances and advancing technology always brings us problems along with its benefits. The taming of fire brought us warmth—and arson. Agriculture brought us a reliable food supply— and a ruined soil. Metallurgy brought us better ploughs—and more dangerous spears.

The advances of science can be applied carelessly and without foresight. It can even be seized upon for purposes that are entirely malevolent.

How is this prevented? How are the possible undesirable side-effects of technological advance foreseen and guarded against?

Again by scientific research. Knowledge may lead us astray, but the answer to that can only be more and better knowledge. Ignorance in the hands of fools has never solved anything.

Thus, when fire was first brought into the home, the smoke must have quickly driven the family out of the home. The possible solution is either to retreat, or advance. It was possible 1) to use fire only outdoors and give up the advantage of an indoor fire, or 2) to devise a chimney.

All throughout history, human beings have chosen the second alternative. Retreat has always been unacceptable to humanity, and when retreat is forced upon a particular society through disaster, human or otherwise, the retreat is the most disastrous consequence of the disaster and later generations refer to it as "a dark age."

So it is now. Humanity stands at the pinnacle (so far) of technological advance, and the problems we face are (as a result) more threatening than any we have ever yet encountered. The population increase, the nuclear weapons, the rapid consumption of resources, the sudden flood of pollution—all of which technology makes possible—threaten to overwhelm us utterly.

But science and technology do not stand still, and we have at hand the very advances that will help us solve these problems. There are two basic scientific technological revolutions already under way that will completely alter the future for the better.

The first is the computer which, with every year, becomes more rapid, more compact, and more versatile. In its essence, the computer is a device to record, manipulate, and retrieve information. The human brain can do it, too, but the computer can do it more quickly; *much* more quickly.

There is no problem that can be solved by recording, manipulating, and retrieving information that a human being can't solve as well as a computer, *given time*. But there is no time to spare, so it comes to this: With computers we can solve problems that, without computers, we can't solve.

The very problems that we now face, if we can solve them at all, we can solve only with the help of computers.

Therefore, all fears about computers, whether justified or not (and I believe them to be unjustified, and even superstitious), are irrelevant. We have no choice but to use them, and to study methods for improving them.

In fact, if we advance our computers far enough, we may be able to foresee the effect of specific technological advances, or combination of advances, well enough and far enough ahead to minimize the very problems that arise through technology. We may learn to avert many of our difficulties to begin with, rather than having to labor to correct them after the fact.

Through computers we may even avoid the dangers that might arise as a consequence of computers.

The second revolution involves space.

Many of our problems arise from the fact that we have filled the Earth. The human population overloads it; human hunger for energy depletes it; human activity poisons it. And we have nowhere to escape.

Don't we?

There are still empty regions, emptier, vaster, richer than any we have yet occupied on Earth. Beyond Earth is space. Out there, there is solar energy for the taking; energy at every wavelength in a perpetual day. There is vacuum, there are high temperatures, energetic radiation, cold temperatures, gravity-free conditions, all of which could be very useful in various industrial and technological procedures.

A quarter of a million miles away there is a huge chunk of real estate called the Moon that is ready for use by us. It has no native life-forms, not even the simplest viruses, to dispute our possession or to raise humanitarian doubts in our hearts.

In space, using lunar material chiefly, we can build solar energy collecting stations to serve as Earth's new never-ending, non-polluting power source. We can build observatories and laboratories and small worlds on which people could live. We can put whole industries into orbit around the Earth.

In fact, we can remove the undesirable side-effects of industrialization from the Earth, not by retreat but by advance. We will still have the benefits of those industries, for they will still exist only a thousand miles away—straight up.

In later years, we can reach and mine the asteroids, spread throughout the solar system, and, eventually, reach out for the stars. In this way, we will not only have a better Earth, but we will also have a never-ending, ever-expanding horizon to maintain the excitement and stimulation of the pioneering spirit.

Both revolutions are in clear swing. The pocket computer invades the home in the 1980s as the television did in the 1950s. The world's governments will have recognized humanity's destiny in space and will be actively beginning to work toward that end.

And Orwell will have been proven wrong!

Unless, of course, the wrongheaded of the world win out, so that scientific research and technological advance slows down and humanity retreats. In that case, civilization will die under the weight of unsolved problems and the consequence will be worse than any Orwell tried to picture.

——— 18 ———

THE BRAIN

Nowadays, we take it entirely for granted that the human brain is the organ that controls thought. We say "He has brains" when we mean that he is intelligent. We tap our temples significantly when we wish to indicate that someone doesn't think clearly. Or else we say, "He has bats in his belfry," meaning that disorderly and unpredictable events take place in the highest portion of the body (the brain), which corresponds, metaphorically, to the highest portion of the church (the belfry), in which bats might literally exist. This is sometimes shortened to a simple "He's bats."

Yet what we take for granted was not necessarily obvious to the ancients. The brain after all does nothing visible. It simply sits there. How different from the heart, which beats constantly all the moments you are alive and no longer beats when you are dead. What's more, the heartbeat races after muscular effort, or when you are stirred by deep emotion of any kind, and it slows during sleep when you seem to be simulating a kind of death.

There is a certain sense, then, in supposing the heart to be the seat of life and emotion. This supposition held sway over so many ages that it remains enshrined in our language. A person who is brave is "lion-hearted," while a coward is "chicken-hearted." If we embolden ourselves to dare a difficult task, we "take heart," and if we suffer a disappointment in love or ambition, we are "broken-hearted." (Needless to say, the heart has nothing to do with any of this.)

If the heart is central to our life, surely that must be so because it pumps blood. A wound that involves the loss of blood weakens us, and, if bad enough, kills us. Blood surges into our face and reddens it during physical exertion or when we are driven to anger or shame. On the other hand it drains from our face, leaving it pale, when we suffer fear or anxiety.

The importance of blood also leaves its mark on our language. When we act under the stress of emotion, we do something "in hot blood." When it is not emotion but calculation that is the spring of our action, we do it "in cold blood." Someone who is commonly emotional is "hot-blooded," someone commonly intellectual is "cold-blooded." Needless to say, the blood remains at the same temperature under all nonpathological conditions.)

Organs particularly rich in blood are also suspected of having much to do with one's state of mind. The liver and spleen are two such organs. Blood is pictured as leaving the liver at moments of fear, just as it leaves the face, supposedly making the dark color of the liver pale; that's why a coward is spoken of as "lily-livered." The word "spleen," on the other hand, refers not only to a blood-filled organ of our body, but also to such emotions as anger and spite. (Needless to say, the liver and spleen have nothing to do with emotions.)

But what about the brain? Does it do *anything?* Aristotle, the most renowned of the ancient thinkers, believed that the brain was designed to cool the heated blood that passed through it. It was merely an air conditioning device, so to speak.

And yet there is one point that might have stirred the suspicions of a careful observer. The abdominal wall contains no bone but consists merely of a tough layer of muscle. The liver and spleen (and other abdominal organs) are thus not guarded very well.

The heart and lungs, which are located in the chest, are better protected, thanks to the bony slats of the rib cage. This seems to indicate that the heart and lungs are more immediately vital to the body than the abdominal organs are. On the other hand, the protection isn't perfect, for a knife can easily slip between the ribs and into the heart,

The brain, on the other hand, is almost totally enclosed by a closely fitting curve of bone. The brain lies hidden inside the strong skull, well-protected from all but the most powerful blow. It is the only organ so thoroughly protected, and surely this must have meaning. Would a mere air conditioning device be so tucked away behind armor when the heart is protected only by slapdash ribs?

This may have been one of the reasons why the ancient Greek anatomist, Herophilus, in the generation after Aristotle, decided that it was the brain that was the seat of intelligence. His opinion, however, did not weigh sufficiently against the overwhelming prestige of Aristotle, whose word was taken as final for nearly two thousand years.

It was dimly understood that the nerves were important, however, and in 1604 an English physician named Thomas Willis wrote the first accurate treatise on the brain and showed that nerves emanated from that organ. That book (only a little over three centuries ago) marked the beginning of the final realization of the brain's importance.

The more scientists studied the brain, the more complex it seemed to be. In its three pounds are packed ten billion nerve cells and nearly a hundred billion smaller supporting cells. No computer we have yet built contains a hundred billion switching units; and if we did build one that large there is no way in which we could as yet compact them into a structure weighing as little as three pounds.

What's more, the "wiring" of the brain is far more complicated than that in any computer. Each nerve cell is connected to many other nerve cells in a complex pattern which allows the tiny electrical currents that mark nerve action to flow in any of a vast number of possible pathways. In comparison, the structure of a computer's units are primitively simple and the patterns of flow easily calculable.

Finally, whereas in a computer the units are switches that are either "on" or "off," the nerve cell units of the brain are themselves magnificently complex objects, each one containing enormous numbers of complicated molecules whose system of functioning is unknown to us but which undoubtedly makes each individual cell more complex than an entire computer is.

The human brain, then, is the most complicated organization of matter that we know. (The dolphin brain might conceivably match it, and there may be superior brains among extraterrestrial intelligences, but we have, as yet, very little knowledge concerning the organization of dolphin brains and none at all concerning those of extraterrestrial intelligences—who might not even exist.) The human brain is certainly more complicated in organization than is a mighty star, which is why we know so much more about stars than about the brain.

The brain is so complex, indeed, and human attempts to understand how it works have, till now, met with such apparently insurmountable hurdles that it seems a fair question to ask whether we can *ever* understand the brain, whether it is *possible* to do so.

After all, we are trying to understand the brain by using the brain. Can something understand itself? Can the brain's complexity comprehend a brain's complexity?

If one human brain were alone involved, these questions would be fair, and might be answered in the negative. However, not one human brain but many are tackling the subject; not one human being but a scientific team that is scattered over the world is doing so. Each researcher may, after years of work, add only a trifling bit to the sum of our neurological knowledge, but all the researchers together are making significant and, in some cases, astonishing progress.

Considering that the human brain, thanks to its intelligence and ingenuity, is the greatest hope of humanity, and that the human brain, thanks to its ability to hate, envy, and desire, is also the greatest danger to humanity—what can conceivably be more important than to understand the various aspects of the brain and possibly to learn how to encourage those that are constructive and to correct those that are destructive?

DARWIN AND
NATURAL SELECTION

The notion of biological evolution is quite old. It began when biologists tried to classify living things. The Greek philosopher Aristotle was among the first to do so, back in the fourth century B.C.

Eventually, in 1737, the Swedish botanist Carolus Linnaeus worked out a system in which living things could be arranged into different kinds (species), similar species could be gathered into groups, and these into groups of similar groups, and so on. It became possible to draw a diagram separating all of life into a few chief branches—each of these into smaller branches—each of these into still smaller branches—until one finally ended with individual species, rather like the individual leaves of a tree.

Imagine that through some magic all we could see of a real tree were its individual leaves distributed in space. Would we suppose that somehow those leaves had just sprung into existence where they were? Surely not! We would suppose they were part of a tree which had grown from a simple shoot, developing branches and sub-branches from which the leaves grew.

In the same way, scientists began to wonder if there might not be a "tree of life" that grew something like an ordinary tree, if present-day species might not have developed from a simpler species, and those from simpler species still, until originally all had developed from one original form of very simple life. This process is called "biological evolution."

Through the 1800s, scientists discovered and studied objects in the rocks that were called "fossils." They had the shapes of bones, teeth, shells, and other objects that had once been alive but had been trapped in rock for millions of years until they had slowly turned into rock themselves.

These fossils were forms of life that were not quite like living species, but they were related to them. The fossils filled in earlier branches in the tree of life and gave hints as to the way in which particular species of life had evolved. For instance, there were horselike animals that lived millions of years ago. They were small to begin with, and they had as many as four hooves on each leg. As time went on, other species were found which were larger and had fewer hooves, until finally the modern horse came into being.

There were other animals that left no descendants, like the magnificent "dinosaurs," huge creatures that were related to modern reptiles (particularly alligators), which all went out of existence, or became "extinct," sixty-five million years ago.

Even though many scientists began to suspect that biological evolution had taken place, it didn't sound very convincing because no one knew *how* it could take place. What could possibly make one species change into another? No one had ever seen a species change. Cats always had kittens, dogs had puppies, and cows had calves. There were never any mistakes.

The first scientist to make a serious attempt to work out the how of evolution was a Frenchman, Jean Baptiste de Lamarck. He thought it arose out of the way in which organisms lived. If an antelope fed on the leaves of trees, it would spend its life stretching its neck to reach leaves higher and higher on the tree. Its neck would grow slightly longer over a lifetime of stretching, and its young would inherit the slightly longer neck. Generation after generation would stretch their necks still further, until finally, after a long, long time, the giraffe would have developed. This process was called "evolution by inheritance of acquired characteristics."

But the theory didn't work. In the first place, acquired characteristics aren't inherited. You might cut the tail off a mouse, but its young will all be born with tails that aren't even shorter than normal. In the second place, how did the giraffe get its splotched coat, which blends in so usefully with the background of splotchy shadows cast by trees, thus hiding it from its enemies? Can the giraffe have *tried* to be splotchier? Of course not.

Then in 1859 an English scientist named Charles Darwin published a book known as *The Origin of Species,* which introduced a real solution to the problem.

Darwin considered that living organisms generally have more young than can possibly be supported by the food supply. If all the baby deer that were born grew up to be adult deer, generation after generation, there would soon be enough deer to strip the trees of vegetation, and all would starve. This doesn't happen, because only a few of the baby deer live to become adult. Most are eaten by other animals. There is competition among the baby deer, so to speak, to see which can remain alive long enough to have baby deer of their own.

Consider this, too. When you study young animals, you find they are not exactly alike. There are always some differences. Some are a little stronger than others, or a little faster-running, or have a color that blends them in a little better with the background and hides them, or whatever. In the competition to grow up safely, some have advantages that work in their favor. In other words, they're the ones that are more likely to grow up and pass their characteristics on to their young. There characteristics, you see, are not acquired, they are *inborn.* They are "natural variations."

Human beings take advantage of the natural variation in their domesticated animals and plants. They select horses that are faster, cows that give more milk, hens that lay more eggs, sheep that have more wool, wheat that grows more ears of grain, and see to it that those in particular give birth to young. In this way, over thousands of years human beings have developed domestic breeds that are far different from the originals—and far better, for human purposes.

Nature does this too. It selects from among the young those that have a better chance, those that are faster and can outrun their enemies, those that are stronger and can beat off their enemies, those that are cleverer and can outwit their enemies, those that have better teeth and can eat more efficiently, and so on.

In this way, horselike animals grew larger and stronger, and developed fewer hooves per limb so as to be more efficient in running. This is through selection, not by people but by nature. It is "evolution by natural selection." Since people work with intelligence, they can produce noticeable changes in a few generations. Nature works hit-and-miss, however. Often the better organism manages to get caught by an enemy through a stroke of bad luck. Evolution by natural selection can require millions of years, therefore, to form new species.

The ingenuity of Darwin's notion of natural selection, and the careful way in which he presented observation and reasoning in this book, convinced some scientists at once. As time passed, it convinced more. Nowadays, scientists generally accept biological evolution on an essentially Darwinian basis. They accept the importance of natural selection as the chief driving force of such evolution.

There were, however, problems from the start, and in the century and a quarter that has passed since Darwin's books, many improvements and advances have been made.

For instance, natural selection depends on inborn variations, but how are these preserved? Suppose that a particular color arrangement is useful as camouflage and that an animal with that color is more likely to survive. It may mate with an animal with a different color arrangement, and if the young have intermediate colors, the advantage will be lost.

In the 1860s, however, an Austrian botanist, Gregor Mendel, experimented with pea plants that showed different characteristics of one sort or another. He crossed one with another and observed the characteristics in the seedlings as they grew. It turned out that characteristics did *not* blend into intermediate forms. Thus, if tall plants were crossed with short ones, some of the seedlings grew tall and some grew short, but none were intermediate.

Mendel published his results, but his paper was ignored. It was not until 1900 that other botanists coming up with similar results looked through scientific journals to see what had been done before and came across Mendel's paper. Mendel had died in 1884, so he never knew that he had founded a new science: "Mendelian genetics."

Mendel had supposed that there were some sort of objects in organisms that controlled the individual physical characteristics of those organisms, and that these objects were passed on from parents to children. In 1879, a German scientist, Walther Flemming, discovered the tiny chromosomes inside the nuclei of cells. Once Mendelian genetics was rediscovered, it was quickly seen that the chromosomes were passed on from parents to young and that this took place in such a way as to account for the manner in which characteristics were inherited. The chromosome was considered to consist of a string of "genes," each of which controlled some particular characteristic.

These genes consisted of large molecules of "nucleic acid" which produced replicas of themselves each time a cell divided. Each new cell thus had the characteristics of those from which it arose.

However, the replica was not always produced perfectly. Tiny accidental changes might be introduced in the molecule. These changes are called "mutations." It is the mutations that produce the differences between one individual and another. It is the mutations that are responsible for the inborn variations among the young that make natural selection possible. Natural selection allows some mutations to flourish and others to die out, and as different mutations survive here and there, new species form.

By 1927, an American scientist, Hermann Muller, showed how one might actually produce mutations by bombarding organisms with x-rays, which change the atomic arrangement in the genes. In 1953 an American, James Watson, and an Englishman, Francis Crick, explained the detailed structure of nucleic acids, and showed how a particular molecule produced its own replica and how it might make a mistake in doing so.

All this strengthened and improved the Darwinian theory of evolution by natural selection.

Meanwhile, since Darwin's time, more and more fossils have been found, and more and more has been learned about the behavior of living organisms and their influences on each other. The actual details of evolution—which organisms descended from which and through what kind of intermediate steps—became better known.

In addition, it was found that natural selection did not always work with mechanical certainty; there were other factors involved.

For instance, chance played a greater part than might have been thought. Where there were small populations of a particular species, it might be that mutations that weren't particularly useful would be established just because a few lucky breaks insured that those individuals possessing those mutations would happen to survive.

In fact, nowadays some scientists, such as Stephen Gould, are thinking in terms of evolution that proceeds very slowly most of the time, but quite quickly under exceptional circumstances.

When there is a huge population of a species, it may be that no mutation can establish itself against the existence of numerous individuals with other mutations. What's more, a few lucky breaks this way or that wouldn't be enough to push evolution in one direction or another. The species might then continue without much in the way of change for many millions of years.

On the other hand, if a rather small population of that species is isolated in a difficult environment, it becomes much more possible that sheer chance will cause some mutations to die out among them altogether, while others survive in considerable numbers. Under such conditions evolution will be faster, and new species may be formed in merely thousands of years.

It is these intervals of rapid change that might be the chief agent for driving evolution forward.

As things stand, then, we can summarize the status of biological evolution as follows:

1. Just about all scientists are convinced that biological evolution has taken place over a period of billions of years, and that all present species, including human beings, have developed from other species that existed earlier.

2. Just about all scientists are convinced that biological evolution has taken place essentially in the manner described by Charles Darwin, and that natural selection among inborn variations is the basic key.

3. Scientists who study evolution nowadays are in deep disagreement on some of the details of the evolutionary machinery, and we cannot yet tell which side will win out in these disputes. However these disputes are resolved will not affect the general acceptance of Darwinian theory, along with its modern improvements, as the basic description of how life developed on Earth.

——— 20 ———

COOL LIGHT

Everything in the Universe emits electromagnetic radiation in a broad band of wavelengths. Cold objects near absolute zero radiate very long radio waves. The Universe as a whole, with a temperature of 3 degrees above absolute zero (–270° C.), produces such radiation, and it reaches us from all directions equally.

As the temperature of an object rises, the peak of radiation shifts toward the shorter wavelengths. Objects at ordinary Earthly temperatures emit considerable energy in the form of the comparatively shortwave infrared. This is true of ourselves, for instance, with our body temperature of 37° C.

If temperature rises still higher, a certain amount of the radiation shifts into the still shorter wave region of visible light. By the time something has been heated to 600° C., enough light is emitted to be detected by our eyes. The object is then "red-hot," red light having the longest waves in the visible region.

As the temperature continues to climb, the emission of light increases, and shifts to shorter wavelengths. At 900° C. an object is bright red; at 1100° C. enough shorter waves are added to give it a yellowish tinge; and at 1500° C. the object is "white-hot."

At all these temperatures, however, the peak is still in the infrared. The amount of light emitted is only a comparatively small fraction of the total radiation. When it is absorbed by our body, *all* the radiation is degraded to heat. That is why, when we put our hand near a flame, or an electric light bulb, we feel heat. A small part of it is from the absorption of the light itself, but most of it is from the longer wave infrared, which we do not see. It is only when a temperature of about 5700° C. is reached—the temperature of the Sun's surface—that the peak of radiation is in the visible light. By then, so much radiation is emitted that the light, when absorbed, produces the sensation of much heat.

Light which is produced as a result of high temperature is "incandescence," from a Latin word meaning "to grow hot." Until recent times, almost all light human beings experienced—sunlight and the light from burning fuel of any kind—was incandescent in nature, and it was always associated with heat.

Light, however, does not have to be produced at high temperatures only. Every atom and every atom combination (the latter is called a "molecule") has a tendency to absorb radiation of a particular wavelength and to emit radiation of a particular wavelength.

The absorption of radiation raises the atoms and molecules of a substance to an "excited" state, one of higher energy than normal. There is a tendency for such an excited atom or molecule to lose the energy rather quickly. Sometimes, it distributes that energy equally among the many atoms and molecules nearby so that the substance merely grows somewhat warmer. Sometimes, it loses it all at once so that the radiation is reflected. Sometimes it loses only part of it and emits radiation that is less energetic (longer in wavelength) than the radiation it absorbs.

As an example, there are some substances that absorb ultraviolet light, which has shorter waves than visible light and cannot be seen (though it is energetic enough to be capable of damaging the eyes), and then emit radiation with longer waves, waves long enough to be seen as visible light. The emission is in only certain wavelengths, not a whole band of them, so that a particular color is seen.

Thus, if various minerals are in the dark and cannot be seen, and if they are then exposed to the radiation from an ultraviolet lamp, which can also not be seen, the minerals will glow in the dark in various colors. Such light, produced at room temperatures, is "luminescence," from a Latin word for "light."

It is usually very weak light, but it is intense enough to affect our retinas and to be seen, *and the light is all there is.* The energy can be converted to visible light with almost one hundred percent efficiency. There isn't a huge quantity of unseen infrared radiation to be absorbed and to add to our sensation of heat. If we were to touch a luminescent mineral, we would be astonished by the lack of heat we felt, for we have been used to associating light with a great deal of heat. We therefore speak of "cold light." However, even this luminescence degrades to a little heat, so in my opinion the phrase "cool light" is preferable.

Sometimes, there is a delay in the emission of visible light after the absorption of ultraviolet. Then, even after the ultraviolet is turned off, the minerals continue to glow for a period of time. This delayed luminescence is called "phosphorescence," from a Greek word for "light," and the minerals capable of displaying this kind of light are called "phosphors." Actually, the difference is not significant from a scientific standpoint, and we might as well allow the term "luminescence" to include "phosphorescence" as well.

Chemists have been studying luminescence and preparing phosphors since the 1600s, but there seemed no practical use for them. The light they produced seemed too weak to be good for anything. And yet luminescent substances were responsible for the revolution in physics, a revolution which began with the discovery of x-rays and radioactivity in the 1890s.

In 1895 the German physicist Wilhelm K. Roentgen was working with cathode rays (speeding electrons) inside evacuated glass vessels. When these electrons struck the glass, their energy excited the glass molecules, which then gave off a faint luminescence. Roentgen wanted to study this luminescence, so he darkened the room and turned on the cathode rays. As it happened, somewhere in the room he had a piece of paper covered with a phosphor, barium platino-cyanide. When he turned on the cathode rays, he caught a glimpse of light out of the corner of his eye. The barium platinocyanide was luminescing strongly. Roentgen investigated and found that when the electrons struck the glass they produced a radiation which penetrated matter and induced the luminescence. He called the radiation x-rays, because he didn't know what it was, at first. It turned out it was like ultraviolet rays, but with even shorter waves.

The study of x-rays at once became all the scientific rage, and in the next year, 1896, a French physicist, Antoine-Henri Becquerel, began to investigate them. Becquerel's father had been interested in luminescent substances, particularly one called potassium uranyl sulfate, which contained uranium atoms, and Becquerel wondered whether this might not give off x-rays when luminescing upon exposure to sunlight. Well, he didn't find x-rays, but he did find other kinds of radiation that had nothing to do with luminescence. In fact, he found that uranium was breaking down and producing gamma rays—which have even shorter waves than x-rays do—together with other kinds of radiation as well.

And thus the new physics began!

In 1910 a practical, everyday use for luminescence was found. A French chemist, George Claude, found that when an electric current was passed through a tube filled with the rare gas neon the neon fluoresced with a bright red light. It would have been unpleasant to try to read by such a light, but if the glass tubes were twisted into the shape of letters, they could be seen (and read) at long distances, and advertisements with "neon lights" began to abound in the cities. Tubes filled with other kinds of gases also fluoresced in various colors, making the advertisements even more colorful.

If an electric current were forced through a tube containing mercury vapor, the vapor luminesced with a light rich in ultraviolet. This was even less useful for reading. However, suppose the inside surface of the tube was coated with a mixture of phosphors that, in combination, luminesced with white light when struck by ultraviolet. Electricity, passing through the mercury vapor, would then produce ultraviolet which would, in turn, produce the white luminescence. After World War II, such "fluorescent lights" became very popular. They produced light with very little infrared, so that the same amount of light could be produced by much less electric current, and the fluorescent lights stayed fairly cool to the touch.

The old cathode rays that made glass fluoresce found a new use, too. The inside of the glass was coated with phosphors that could luminesce with white light or with different colors. If the stream of electrons moves progressively across a screen, becoming weaker and stronger according to the lights and shadows of a scene as photographed by a camera, a picture is painted in either black and white, or color, depending on the phosphors. The picture will move as the real object moves before the camera. The cathode rays and the phosphors make up a television tube, in other words.

The laser, first devised in 1960, is a still more recent technological development that involves luminescence. Here the light can be concentrated till it is brighter and hotter than the Sun. Luminescence is not *always* cool light.

Luminescence can be produced by the absorption of almost any kind of energy. Electric currents, ultraviolet radiation, and x-rays have been mentioned, but chemical changes can also produce luminescence.

Many chemical changes produce considerable amounts of energy. Thus, the burning of fuel—wood, coal, oil, wax—produces so much energy that almost our entire technology is run by such combustion. Most of the energy of chemical change appears in the form of heat, together with the kind of light produced by incandescence.

Comparatively small amounts of the energy of chemical reactions can serve to produce excited atoms which then lose their energy in the form of light. This "chemiluminescence" is generally so small we can't detect it. In a very few cases, however, it can be seen. If hydrogen peroxide is added to a chemical known as "luminol," the chemical glows with a blue-green fluorescence. Even in this case, only about one percent of the energy of the chemical reaction is turned into light (though in recent years some chemical reactions have been discovered in which up to nearly a quarter of the energy is turned into light).

For real efficiency, though, we have to turn to life. In living tissue there are many chemical reactions that take place under the influence of certain complicated molecules known as "enzymes." The enzyme molecules possess specialized surfaces on which the chemical changes that usually take place very slowly take place very rapidly indeed.

Suppose there is a particular chemical reaction which, if it takes place in a certain way, will produce a luminescence. If an appropriate enzyme is present which "catalyzes" this reaction—that is, makes it take place very quickly—all the molecules will undergo that particular change before any of them have time to undergo any other change. The result is that just about all the energy of the reaction will be converted into luminescence, and the efficiency of the light production will be nearly one hundred percent. Such luminescence in living tissue is "bioluminescence."

The best known example of such a chemical reaction is that involving a chemical known as "luciferin" (yes, from "lucifer," meaning "light-bearing," a name given to the morning star because it heralded the rising of the Sun, and, for Biblical reasons, to the devil). In the presence of the enzyme "luciferase," luciferin combines with oxygen to yield energy. A single molecule of luciferin produces a tiny flash of light, too small to see, but many millions of them, all flashing, turn a firefly's abdomen into a small lantern. After the flash, the firefly is dark while the luciferin reforms, then it flashes again.

It is a tiny light, but quite visible at night. Anyone catching a firefly, who knows nothing about luminescence, will be astonished at an insect that gives off light, without heat—that seems to be burning, without harming itself.

Luciferin has been extracted from fireflies by the hundreds of thousands, and, in the presence of the proper enzyme and certain other substances that participate in the overall rection, the firefly's lantern can be duplicated in the test tube, and the fluid within the test tube will glow yellow in the dark.

Why do fireflies glow? So they can see? Clearly not, for there are countless kinds of night-flying insects (and other animals) who do not require bioluminescence to get around in the dark.

Rather, the firefly's little light helps solve a great problem that all forms of life have—how one individual can recognize another individual of the same species so that they can mate and keep that species alive. There are many strategies that are made use of by different species, involving sounds, odors, movements, and so on. As for the firefly, it uses flashes of light as a signal to possible partners, and it works very well.

Bioluminescence is uncommon on land, however. Fireflies are a rare exception. Bioluminescence is not found in any of the land vertebrates: amphibia, reptiles, birds, or mammals. Nor is it found among the green plants, though there are some luminescent mushrooms. It is also almost never found in life in fresh water. Where bioluminescence comes most nearly into its own, however, is among animals that live in the sea.

This is, perhaps, to be expected, for sunlight does not penetrate beyond the ocean's inconsiderable surface layer, and most animals of the sea dwell in utter darkness. There, bioluminescence can have its uses, and there is, in consequence, so much luminescence in the ocean depths that it is actually useful for species to have functioning eyes and to be able to detect light. The result is threefold—

1. Members of a particular species can recognize each other and remain together for mating and other purposes.

2. Predators can follow light to possible sources of food, or use light as a way of getting it.

3. Organisms can use light to mislead predators and save themselves.

The best known examples of predators that use light as a lure are the various species of angler fishes. They are so called because attached to their heads, just before the mouth, is a little bioluminescent object. Any small living thing that is attracted to it and approaches is swallowed. In some cases the bioluminescent bulb is shaped like a worm and wriggles. A small fish approaches to feed and is fed upon instead. In other fish it is the tongue itself that luminesces and resembles a worm so that prey actually swim into the mouth.

Flashlight fish use light for protection. They can cover up bioluminescent patches under their eyes at will. In this way, they can glow, or not glow. A school of flashlight fish will take up a zigzag path, glowing when they zig, but not glowing when they zag. Predator animals, following the zigs eagerly, find themselves missing the mark, as the hoped for prey moves darkly in another direction.

The pony fish, living in the upper regions of the ocean where there is still some light from above, has a faint glow all over its lower surface. Predators at a greater depth, looking up, fail to see the pony fish against the faint light coming from the air. Bioluminescence becomes a kind of protective coloration.

In the upper layers of the ocean, squids and octopuses can squirt dark "ink" in which they can hide or which can mislead pursuers. Ink would do no good in the black depths, but there are some octopuses and other crustaceans that can, when pursued, expel a cloud of bioluminescent matter in one direction while they flee in another. The pursuer is liable to follow the light and miss the prey.

Then, too, there are species of fish in which only the males have a bioluminescent spot. When a predator approaches a school of these fish, the males flee in all directions, but the females don't move. The predators follow the males and catch a few, but the biologically more important females remain untouched.

There are large numbers of protozoa and bacteria that are luminescent. In the days before refrigeration became common, butchers' meat would glow faintly in the dark after a while because of luminescent bacteria infesting the surface. (Such bacteria are apparently never harmful.)

The sea itself contains one-celled creatures that luminesce whenever disturbed. Ships, moving through tropical waters, sometimes leave a glowing wake behind them, a phenomenon sailors and fishermen used to call "the burning of the sea."

It is these tiny creatures that are sometimes responsible for the bioluminescence of larger animals. The flashlight fish and the pony fish, for instance, are infested with colonies of luminescent bacteria in those parts of their body which glow. The infestation is beneficial for both parties (a phenomenon called "symbiosis") since the fish gets the light and the bacteria have a steady source of food from the fish's bloodstream.

But why should one-celled animals luminesce? There seems no obvious use for such luminescence. For every species that luminesces, there are other closely related species that do not luminesce and that get along very well.

It seems quite clear that bioluminescence evolved, to begin with, in one-celled creatures, perhaps a billion years ago or more, and, apparently, on several different occasions. Surely, there must have been a use for it.

One possibility is that it was a way of handling oxygen. Oxygen is a very active and potentially poisonous substance. Life forms have developed methods for using oxygen to combine with food to produce energy, but to begin with, oxygen may have been an embarrassment that had to be gotten rid of. All systems of bioluminescence are produced by the combination of some form of luciferin (they differ in chemical detail from species to species) and oxygen.

When *any* chemical in the body combines with oxygen, energy is produced, and when the energy is produced in excess, it *must* be gotten rid of. (Failure of human beings to do so efficiently on hot, muggy days produces the discomfort we are all familiar with.) The value of luciferin over other systems is that the energy is at once radiated away in the form of light.

And there we have the history of luminescence. From glowing minerals, and from bacteria which, in the very dawn of life, were getting rid of dangerous oxygen and waste heat—to the glamor of modern technology with its television and laser beams.

— 21 —
HALLEY'S COMET—
DESTINATION SPACE

The year 1985-1986 was the time of Halley Hysteria. Everyone was whooping it up for the return of the comet. No one seemed to pay attention to the fact that it was going to be a bad show, that the comet would never come close to the Earth, and that it would be angled so as to be the most visible in the southern hemisphere.

Newsweek asked me to do a big article on the comet for an advertising supplement they were expecting huge things from. I was doubtful, but I am an easy man to talk into doing an article. In the end, I agreed. But, as you might expect, the advertisers didn't turn up, and Newsweek ran a small supplement which included only a tiny portion of the article.

I should have followed my own reasoning, but there is always an element of risk in free-lance writing and I am used to it. In any case, here is the article as I wrote it, in full.

Throughout history, comets have been frightening objects. Ordinary stars move in stately, never-changing circles about the sky. The Sun, the Moon, and the planets move in more complicated fashion, but even those motions have a regularity about them. Five thousand years ago, the Sumerians studied these more complicated motions and began the process of analyzing them to the point of being able to predict future positions. Sky-watchers have continued and improved the process ever since.

The comets, on the other hand, moved with complete irregularity. They appeared suddenly, without warning; they followed unpredictable paths across the sky; and then they vanished. Another might come a month later, or not for years.

Irregularity was frightening. Once the paths of the planets were worked out, every little movement was given an astrological meaning. Their future positions were taken to predict forthcoming events in the lives of nations and people. An irregular object in the sky meant an irregular event, something out of the ordinary.

The very appearance of the comet was daunting. The stars and planets were dots of light, or circles (or parts of circles) in the case of the Sun and Moon. Comets, on the other hand, were hazy objects of irregular and changing outline, possessing a long filmy tail. It was easy to picture the comet as a sword with a long curved blade, or as a fleeing woman with unbound hair streaming behind her. (The very word "comet" is from the Greek for "hair.") Either way it seemed to presage death and sorrow. No one doubted that the coming of a comet was the gods' way of foretelling disaster, so that the weeks in which it shone in the sky left human beings weak with continuing terror.

Only slowly, in modern times, did comets become objects of scientific interest rather than horrified panic. It was not till 1472 that a German astronomer, named Regiomontanus, reacted with something other than fear and actually plotted the position of a comet against the stars, night after night.

A comet in 1531 was studied by an Italian astronomer, Girolamo Fracastoro, and he pointed out that the tail always pointed away from the Sun. Then in 1577 a Danish astronomer, Tycho Brahe, tried to determine the distance from Earth of a comet that appeared that year. He couldn't get an exact figure, but he showed that it had to be much farther than the Moon. A comet, therefore, was an astronomical object, and not part of the atmosphere as the ancient Greeks had thought.

By then, astronomers were coming to believe that the planets, including the Earth itself, revolved in fixed paths, or "orbits," about the Sun. Did that mean that comets revolved about the Sun, too? There was, however, no good way of calculating orbits until 1687, when the English scientist Isaac Newton worked out the mathematical details of the law of universal gravitation.

Newton had a younger friend, Edmund Halley, who had carefully observed a comet in 1682, and plotted its movement across the sky. He noted that the path it followed was identical to that followed by comets that had appeared in 1607, 1531, and 1456 (according to astronomical records). He noticed also that there was just about seventy-six years or so between these dates. Could those paths have been taken by one comet that returned periodically? Using Newton's mathematics, Halley spent years calculating the comet's orbit. By 1705 he was finished, and announced the shape of the orbit (a very long ellipse, with the Sun near one end), predicting the comet would return in 1758.

Halley didn't live to see the return, but the comet came back and it has been known as "Halley's Comet" or, more recently, as "Comet Halley," ever since. After 1758, it returned in 1835 and 1910. It arrived once more in 1985-1986.

Halley's Comet remains the most famous of all comets, but it is not the only one, of course. Many comets with orbits that carry them back to Earth's vicinity periodically are now known. One of them, Encke's Comet, has the shortest orbit, returning every three and one-third years. These days, a comet is spotted by astronomers every two or three weeks or so. Most of these comets are very faint, however.

Every once in a while, an extremely bright comet showed up in the sky. There were half a dozen such in the 1800s, but no really bright one since 1910, when Halley's Comet last appeared. Even this year's Halley's Comet is not very bright in Earth's sky because it is passing us at a greater distance than usual.

Really bright comets are "new comets" with such elongated orbits that they don't approach earth oftener than once in every million years or so.

Nowadays, astronomers believe that there is a vast sphere of comets enclosing the solar system very far out from the Sun, perhaps a thousand times as far as the farthest planet. There may be a hundred billion comets spread over this sphere, each five or ten miles across. The comets are too small to see, of course, and are so spread out that they don't in any way block our vision of the Universe beyond.

Every once in a while, either because of collisions, or because of the pull of some of the nearer stars, a comet moves out of its distant orbit and falls toward the Sun. It wheels about the Sun and returns to its original spot, doing so over and over, taking millions of years for each return. Eventually, it may pass a planet which "captures" it, pulling it into a much smaller orbit about the Sun. The comet then returns every few years or decades.

Far from the Sun, the comet is extremely cold and is a glob of icy substances (mostly ordinary frozen water) with rocky dust and gravel spread through it. (This is the "dirty snowball" theory.) As the comet approaches the Sun, the ice melts and vaporizes, liberating the dust which lifts off the comet and reflects sunlight hazily. The solar wind (which consists of speeding subatomic particles shot out of the Sun in all directions) sweeps the haze outward away from the Sun, stretching it out to a long tail if there is enough of it.

Each time a comet passes the Sun it loses much of its ice and dust, which never returns. Some of the gravel remains, however, and eventually forms a dark and thickening layer about the ice which eventually no longer has a chance to melt. The comet is then dead and continues to circle the Sun as though it were an asteroid. Sometimes there is so little gravel that all the ice melts and the comet vanishes altogether. Astronomers have witnessed such vanishings. Halley's Comet is large enough to survive a great number of turns before dying. It will, however, grow gradually dimmer with the centuries.

Comets are of particular interest to astronomers because they may illuminate the far past.

The solar system is thought to have formed into more or less its present shape about 4.6 billion years ago. Before that it was a vast cloud of dust and gas that for some reason (possibly the explosion of a nearby giant star, setting off a pressure wave in the cloud) began to condense and to whirl faster and faster as it grew smaller.

The central portion of the cloud, which collected 99.9 percent of the mass of the present solar system, was so large and massive that its core underwent the kind of pressure and temperature increase that ignited nuclear fusion reactions. It became the Sun. The 0.1 percent of the matter outside the Sun condensed into planets, satellites, asteroids, and so on. The details of the condensation have been carefully worked out by astronomers, but, of course, many of the conclusions are strictly tentative. There isn't enough data concerning the chemical makeup of the original cloud, for one thing, to make astronomers feel very sure of their calculations and conjectures.

We can't tell very much about the original cloud from most of the objects in the solar system today. The Sun has been undergoing nuclear changes for billions of years, so we can't be sure of its original makeup. The planets have been undergoing chemical separations so that we can't understand their makeup unless we know what they're like far beneath their surface and learn the details of the chemistry of their interiors.

Asteroids and meteors might be more useful because their interiors are less mysterious. In fact, we can actually analyze some meteors we pick up after they have fallen to Earth. However, these small bodies have been exposed to the heat of the Sun and to the solar wind for so long that astronomers feel that they, too, have undergone obscuring changes.

But what about the comets? They were formed 4.6 billion years ago as the rest of the solar system was, but the comets formed about five to ten trillion miles from the Sun, and have stayed there ever since. Halley's Comet has been in its present orbit only some thousands of years at most, or it would be dead by now. At the distance of the comet cloud, the Sun seems no more than a bright star and does not perceptibly affect the comets. The comets therefore are made up of a mixture of ices, dust, and gravel, that represents the original pre-system cloud (minus the very light gases of hydrogen and helium).

If we could study a comet in chemical detail, we might well learn a great deal more about the original cloud than we now know, and we would be able, perhaps, to work out the details of the origin of the solar system, including the origin of the Earth, much better than we can now. And if we knew the details of the early Earth better than we do, we might understand more nearly how life began in the first billion years of Earth's existence—and therefore a little more about ourselves.

All from a comet.

Of course, until now there has been no way of studying comets. We have never even seen the solid core of a comet. Any comet that is close enough for us to study through a telescope with some hope of getting some detail concerning its surface is also close enough to the Sun to develop a haze that completely obscures its surface.

What we need is a space probe that will approach Halley's Comet as it obligingly nears a planet only now populated by a species technologically advanced enough to build a space probe.

The United States, unfortunately, has chosen not to join in the venture out of a reluctance to spend money on the project. It has, to be sure, sent a probe that was already in space through the tail of a small comet named Giacobini-Zinner after its discoverers. The probe was not adapted for the purpose, however, and could only send back some minimal (but useful) information concerning the tail.

Four true comet probes have been launched toward Halley's Comet, however—two by the Soviet Union, one by Japan, and one by a consortium of Western European nations.

The most ambitious of these is the European probe, which has been named "Giotto" (pronounced "JOT-oh"). The name seems odd, but it makes perfect sense.

In 1301, Halley's Comet made one of its appearances and it must have been observed by the Italian painter Giotto di Bondone. In 1304 Giotto (he is always known by his first name) painted "The Adoration of the Magi," a beautiful masterpiece depicting the three magi worshipping the infant Jesus. Above the manger, Giotto drew his version of the Star of Bethlehem and he made it a comet, undoubtedly drawing on his memory of the appearance of Halley's Comet three years earlier. It is the first *realistic* drawing of a comet in history, and that seems to be reason enough to honor the painter by giving his name to the probe.

If all goes well, Giotto will travel right through the cloud of haze surrounding the solid core of the comet and pass within 500 miles of its icy surface. This may be a suicide mission, for it is quite possible that the dust in the cloud may scour the probe or puncture it and put it out of commission, but the hope is that it will survive long enough to send back valuable information.

Of course, even if Giotto survives, the haze will be sure to obscure the comet's surface. However, there are techniques for penetrating haze. We have looked through the cloud layer on Venus to map the surface beneath, so the comet's haze should offer no insuperable difficulty.

And when Halley's Comet returns again, in 2062, it may be that automated probes will be able to scoop up some of the cometary ice and analyze it on the spot. Or robots may land on the surface and report. Or, perhaps, even men— properly protected.

Suppose we reverse the position and imagine intelligent beings on Halley's Comet studying Earth each time they pass it at seventy-six-year intervals. What would they see? Halley's Comet, has, of course, been passing Earth since prehistoric times, but the first report we have of a comet sighting that *might* be Halley's Comet dates back to 467 B.C. Let's start there:

1. 467 B.C. There are civilizations in China, India, and the Middle East. Ancient Greece is in its Golden Age. Pericles rules Athens and Socrates is a young man.

2. 391 B.C. The little town of Rome has been sacked by the Gauls. There is no indication yet of its future greatness.

3. 315 B.C. Alexander the Great has died after conquering the Persian Empire, and his generals are fighting over fragments of his conquests. The Museum at Alexandria has been founded.

4. 240 B.C. The greatest ancient scientist, Archimedes, is at work, and Buddhism is spreading over India.

5. 163 B.C. Rome, having defeated Carthage, now controls the western

Mediterranean. The Jews have rebelled against Syria and are setting up the Maccabean kingdom.

6. 87 B.C. Rome has now extended its control into the eastern Mediterranean, but its generals are beginning to fight each other, and there is the confusion of civil war.

7. 12 B.C. Augustus has ended the civil wars, turned Rome into an Empire, and rules as its first (and best) Emperor. In Judea, now a seething Roman province, Jesus will soon be born.

8. 66. Rome (under Nero) and China (under the Han dynasty) are both large and prosperous. Judea has rebelled and will soon be crushed. The Jews will be deprived of a homeland for nineteen centuries.

9. 141. The Roman Empire under its "Good Emperors" is at the peak of its size, culture, and prosperity.

10. 218. Roman citizenship has now been extended to all freemen in the Empire, but civil wars and economic dislocations are once again beginning to undermine the realm.

11. 295. Christianity has been spreading rapidly through the Roman Empire and it will not be long before it will become the official religion.

12. 374. The Germanic tribes and the Asian Huns are becoming more and more of a threat. Rome is maintaining itself with increasing difficulty.

13. 451. The Huns under Atilla are devastating the west and the Roman Empire is falling apart.

14. 530. The Eastern Empire under Justinian is strong. Constantinople is the greatest city in Europe.

15. 607. The Persian Empire nearly (but not quite) destroys the Eastern Empire. In Arabia, a new religion, Islam, is being founded.

16. 684. In a rapid sweep, Islamic forces have conquered western Asia and northern Africa and have threatened Constantinople itself.

17. 760. Islamic forces have invaded Spain and France but have been stopped by the Franks, who are now the strongest force in Western Europe and will soon be ruled by Charlemagne.

18. 837. Charlemagne is dead and his Empire is breaking up. The Vikings are harrying European shores.

19. 912. The Vikings have founded Normandy in France, and Saxon England will soon be at its peak under Alfred the Great.

20. 989. France has a new king who founds a line that will rule for eight centuries. Russia is converted to Christianity.

21. 1066. William of Normandy invades England, defeats the Saxons, and becomes William the Conqueror.

22. 1145. The Crusades are in full cry, but their success has passed its peak and the Islamic forces are counterattacking.

23. 1222. The Mongols under Genghis Khan are sweeping across Western Asia and into Eastern Europe. In England, the Magna Carta was signed a few years earlier.

24. 1301. The Mongol Empire, the largest land empire ever, reaches its peak. The medieval Papacy reaches its peak. Both collapse soon after.

25. 1378. The world has survived the Black Death. England and France are engaged in the Hundred Years War. Tamerlane is conquering Western Asia. The Renaissance is at its peak in Italy.

26. 1456. The Islamic Turks have conquered Constantinople and Europe is terrified. The Portuguese are working their way around Africa to bypass the Turks.

27. 1531. The Ottoman Empire reaches its peak when it lays siege to Vienna (and fails). The Protestant Reformation has taken place. Columbus has discovered America, and Spain and Portugal are both establishing vast overseas empires. European domination of the world is beginning and will last four centuries.

28. 1607. England has become an important power, and its first colony in what is now the United States is established. Spain has declined and France is the dominant force in Europe. Modern science is beginning with Galileo.

29. 1682. France is reaching its peak under Louis XIV, while Russia is becoming an important power under Peter the Great. The Age of Reason is launched with Isaac Newton.

30. 1759. Great Britain is defeating France and establishing itself in North America and India as the dominating power of the world. Prussia, under Frederick the Great, has become a strong power. The Industrial Revolution is about to begin.

31. 1835. The Industrial Revolution is proceeding. Steamships and steam locomotives revolutionize transportation. The United States is independent. The French Revolution and Napoleon have convulsed Europe.

32. 1910. The automobile, the airplane, radio, and motion pictures are all with us now. European power is at its peak as it faces the coming debacle of World War I.

33. 1986. Europe is in decline, its empires gone. The Soviet Union and the United States face each other, with world dominion or world destruction the apparent alternatives.

Let us go into greater detail concerning the two twentieth century appearances. Through most of the returns of Halley's Comet, everyday life on Earth did not change much. Even as late as 1759 the world was agricultural, and technology had advanced very little.

By 1837, however, Great Britain and northwestern Europe were industrializing quickly, and by 1910 industrialization had reached even higher peaks in Germany and the United States and was beginning to penetrate Russia and Japan.

The comet's return in 1910 saw the industrialized nations in the golden age of the railroad, but automobiles were moving along the roads, and airplanes were flying across the English Channel. Telegraphy and cables were in their prime, but radio was already spanning the Atlantic Ocean. While at the previous return the world was being "steamed," in 1910 it was "electrified," with the telephone and electric light as developing marvels.

The greatest revolution in science since Newton's time had just taken place. In 1895 x-rays had been discovered, and in 1896 radioactivity was. The atom was found to have a structure and subatomic particles were recognized. Einstein's theory of special relativity was announced in 1905, and it was soon to be extended further and even more magnificently to the theory of general relativity. Planck had discovered the quantum nature of heat in 1900, Einstein extended it to the atom, and soon a galaxy of physicists were to found quantum mechanics.

In astronomy data was being collected that would lead to the notion of the expanding Universe, and pretty soon it would become apparent that the Galaxy was not the Universe, but only one of billions of galaxies.

In biology genetics had been founded, the importance of chromosomes understood, and genes as the units of inheritance were beginning to be talked about. The fact of evolution had been well-established for half a century and was disputed then (as today) only by the superstitious.

In medicine the germ theory of disease had led to the development of vaccines and antitoxins so that infectious disease was being brought under control and the life expectancy was lengthening. Anesthetics and antisepsis had converted surgery from a killing torture chamber to a lifesaver. The recent discovery of vitamins and hormones added new dimensions to nutrition and medicine.

The world was becoming smaller as science became international. Travel, the postal service, news agencies, and weather forecasting all spread out to encompass continents and oceans. The North Pole had been reached and the South Pole was about to be.

There is no question but that 1910 would be recognized today as representing an already modern society in science and technology, but how far we have advanced since!

The last seventy-five years have seen so much and such varied technological advance that we can only touch a few of the highlights.

In astronomy there was a tremendous revolution that began in 1931, when an American radio engineer, Karl Jansky, was attempting to track down the source of static that interfered with radio reception. He came across very faint, steady noises which, by a process of elimination, he decided were radio waves coming from the sky. At first he thought they came from the Sun, but day by day the source moved relative to the Sun, and by 1933 it was clear they came from the direction of the center of the Galaxy.

Hearing of this, a radio ham named Grote Reber in 1937 built a parabolic dish designed to receive the cosmic radio waves. It was the first "radio telescope" and he used it to locate a number of radio sources in the sky.

Nothing much more could be done because astronomers lacked the necessary devices to handle the radio waves. During World War II, however, radar was developed. It made use of beams of radio waves that were emitted, reflected from objects, and detected, thus making it possible to detect distant airplanes in flight, for instance, and estimate their distance. This stemmed from the work done by the Scottish physicist Robert A. Watson-Watt in 1935.

After World War II, the new techniques of radar, or "radio astronomy," were quickly developed. With larger and more delicate receivers, the new telescopes could penetrate farther than ordinary telescopes and could see smaller objects. Wholly new phenomena were discovered. In 1963 a Dutch-American astronomer named Maarten Schmidt discovered quasars, the most distant objects known, up to 10 billion light-years away. They were tiny, not much larger than the solar system, but shone with the light of a hundred galaxies. In 1969 the British astronomer Anthony Hewish discovered pulsars, tiny stars as massive as the Sun, but only eight or so miles across and capable of rotating in fractions of a second. Exploding galaxies and even the faint far echo of the big bang, with which the Universe began about 15 billion years ago, were discovered.

Before radar was invented, the French physicist Paul Langevin was able to do much the same in 1917 with short-wave sound. The device was called "sonar." After World War I, ships began to use the sound reflections to study the ocean floor in detail for the very first time.

By 1925 it was found that a huge ocean range ran down the center of the Atlantic Ocean; eventually it was found to snake through all the oceans. In 1953 American geologists William M. Ewing and Bruce C. Heezen discovered that a deep canyon went down the center of this ridge, and this marked the beginning of the discovery that the Earth's crust was split up into a number of "plates." Slowly these plates move, in some places away from each other, splitting continents and widening oceans. In others they move toward each other, forming mountain ranges, or move one under another to form ocean deeps. Earthquakes, volcanoes, and continental drift (first suggested by a German geologist, Alfred L. Wegener, as early as 1912) all began to make sense in the light of the new science of "plate tectonics."

In 1910 the only subatomic particles known were the electron and the proton. In 1932, however, the English physicist, James Chadwick, discovered the neutron. Since the neutron had no electric charge, it would not be replled by the atomic nucleus, and it would serve as a particularly effective bombarding agent.

The Italian physicist Enrico Fermi at once began to use it to bombard nuclei of various elements. Among the elements he bombarded was uranium, but here his results seemed confusing.

The German physicist Otto Hahn and his coworker, the Austrian Lise Meitner, continued the investigation of bombarded uranium, and in 1939 Meitner (having fled the Nazis) suggested that what happened was that the uranium atom underwent "fission," splitting in two.

In doing so, it liberated additional neutrons that split other uranium atoms, and so on, starting (under proper conditions) a vast "chain reaction." The world was then on the brink of war, and deep secrecy was imposed on such research by the United States. By 1945 a fission chain reaction had been made the basis of an "atomic bomb." Two were exploded over Japan to end World War II.

Meanwhile, as early as 1915, the American chemist William D. Harkins had pointed out that the fusion of the tiny nuclei of hydrogen into the somewhat larger helium nuclei produced an unusually high yield of energy. In 1938 the German-American physicist Hans A. Bethe showed that such hydrogen fusion was the source of the Sun's energy, and presumably that of the stars generally.

After World War II efforts were made by the United States to devise a fusion bomb (the so-called hydrogen bomb), and in 1952 this was achieved, thanks, in part, to the work of the Hungarian-American physicist Edward Teller. The Soviet Union labored to match the United States, first with a fission bomb of its own and then with a fusion bomb. In fact, to date, neither nation has achieved a decisive superiority, and the world is now faced with two opposing powers, each disposing of the capacity to destroy civilization in an hour.

Nuclear power has its more benign side. Uranium fission has been brought under control so that fission power stations can produce electrical energy for peaceful purposes. Hydrogen fusion has not yet been brought under control but may be someday.

Then, too, fission reactions can be used to produce radioactive isotopes in quantity. These isotopes have the chemical properties of ordinary elements but can easily be detected in the body. By being added to the ordinary elements they can be used to work out in detail chemical changes that take place in the body and have vastly increased our knowledge of biochemistry.

Indeed, the ability of chemists to study the large molecules in tissue in great detail, through a number of techniques, has produced the science of "molecular biology." One crucial advance, by the English physicist Francis H. C. Crick and the American chemist James D. Watson in 1953, was the working out of the detailed structure of DNA, which controls heredity and seems to be the key molecule of life.

Since then, other chemists have learned techniques for splitting the DNA molecule in chosen places and putting it together again. In this way, defective genes might be corrected, and new varieties of genes might be produced. Otherwise incurable diseases might be treated, new kinds of life might be created with novel properties, and so on. We stand on the brink of a mighty development of "genetic engineering," or "biotechnology."

Scientific changes can lead to startling changes that might seem at first blush to have nothing to do with science. In the 1950s, for instance, American biologist Gregory G. Pincus developed an artificial hormone that had contraceptive properties. Women who took this "pill" regularly were freed of the danger of pregnancy and could approach sex in as carefree a fashion as men. hhThis, more than anything else, powered the feminist movement and the sexual revolution.

In the 1920s, quantum mechanics was developed by such men as German physicist Werner Heisenberg and Austrian physicist Erwin Schrodinger. Their research was extended to chemistry by American chemist Linus C. Pauling in 1939.

Quantum mechanics was enormously successful, guiding scientists to an understanding of many details of theoretical physics, and helping refine the technology of devices that could carefully control streams of electrons ("electronics").

By 1910 radio was already known, and the "radio tube" was coming into use as a way of controlling the electron stream, but of course matters didn't stop there.

The transmission of sight as well as sound was under investigation by the 1920s, and in 1938 the first practical television camera was invented by Russian-American physicist Vladimir K. Zworykin. After World War II television became sufficiently reliable and cheap to penetrate the home, and it rapidly grew to be the prime source of family entertainment. Quickly, it took over news delivery and grew to dominate the political process.

Streams of electrons could be focused in the same way light rays were and the first electron microscope was devised by German physicists Ernst Ruska and Max Knoll in 1932. It was rapidly improved and became the vehicle whereby biologists could look inside the cell and see enormous detail. The fine structure of even tiny viruses and DNA molecules could be made out.

Light itself was produced in new forms. A suggestion Einstein made in 1917 was studied, and in 1953 American physicist Charles H. Townes built a device which came to be called a "maser," which produced a beam of short radio waves, all of which kept in perfect step. This was called "coherent radiation" because it could be focused far more sharply than ordinary radiation could.

In 1960 this device was adapted to produce a coherent beam of visible light by American physicist Theodore H. Maiman. This was called a "laser," which turned out to have numerous applications. It could be used in surgery, or as a drill puncturing its tightly focused way through any metal. It could carry information far more efficiently than electricity could, so that tiny glass fibers carrying laser light are more and more replacing expensive copper wires carrying electricity. It can even be used as a "death ray," and as such it is an essential part of the plans of the Reagan administration to set up a "Star Wars" weapon in space.

Then, too, the radio tubes were replaced as controllers of the electron stream. They were bulky, fragile, required considerable energy, took time to warm up, and had a tendency to leak and black out. In 1948 English-American physicist William B. Shockley produced a small device made of metals such as silicon or germanium, with small amounts of impurity added, that could do the work of radio tubes. It was called a "transistor" and was tiny, solid, sturdy, used little energy, and required no warming.

Any electronic device using tubes could be "transistorized" and become smaller, cheaper, and more reliable. Transistorized radios, in particular, with batteries for power, became widespread. In the third world, such radios reached and united the people and gave them a national consciousness they might never have had otherwise.

Calculating machines of one sort or another had existed for centuries, but in 1946 two American engineers, John P. Eckert and John W. Mauchly, built the first electronic computer. Equipped with nineteen thousand vacuum tubes, it could solve complicated mathematical problems more quickly and reliably than any other device that had hitherto existed.

Computers evolved with enormous rapidity, and as they were transistorized, they grew smaller, cheaper, *and* more capable. Transistors themselves were "miniaturized." Whole combinations of transistorized circuits were developed on a single "chip." In the mid-1970s, "microchips" were invented on which circuits were all but microscopic. Vest pocket computers could be bought for less than a hundred dollars that, powered by small batteries, could do more than the giant computers of thirty years before.

Computerized machines capable of performing repetitious work on assembly lines, more accurately and more tirelessly than human beings could, came into being. These were the first "robots." Work is now being done to make robots more versatile and fitted for home use.

Although there were airplanes in 1910, they were fragile, slow-moving devices. With the years they were made larger, sturdier, faster, and capable of carrying greater weights, but they remained minor methods of transportation through World War II. In 1939, however, an Englishman, Frank Whittle, flew a "jetplane," one that depended for propulsion, not on a propeller, but on a jet of fuel exhaust, using the rocket principle of action-and-reaction.

In 1947 an American jetplane piloted by Charles E. Yeager surpassed the speed of sound, and in the 1950s commercial flights by the jetplane became more common until, very soon, single planes carrying hundreds of passengers could go from any airport on Earth to any other in a matter of hours. Humanity had reached a new pitch of mobility.

Jetplanes carry fuel, but they depend upon the oxygen in the air to oxidize the fuel and produce the jet of hot exhaust gases. In 1926, however, the American physicist Robert H. Goddard had fired off the first rocket carrying both liquid fuel and liquid oxygen. This first rocket was tiny, but Goddard built larger ones in New Mexico and developed methods for controlling and steering them. During World War II, German rocketry advanced under the leadership of Wernher Von Braun; his V-2 rockets, the first capable of reaching extreme heights, were used by the Nazi regime to bombard London.

In 1957 the Soviet Union put the first rocket in orbit about the Earth, and in 1961 Soviet cosmonaut Yuri A. Gagarin was the first man in orbit. In 1969, the American astronaut Neil A. Armstrong was the first man to stand on the Moon.

Advance has been rapid. Unmanned rockets have sent back photographs from the surfaces of Venus and Mars, and from the near neighborhood of Mercury, Jupiter, Saturn, and Uranus. Soviet cosmonauts have remained in space for up to eight months at a stretch, and American astronauts making use of reusable space shuttles have been flying through near space routinely, performing a variety of tasks, from rescuing malfunctioning satellites to testing techniques for putting together structures in space.

How enormously, then, has humanity advanced in technology in the seventy-six years that mark the gap between the most recent two appearances of Halley's Comet.

Hindsight is easier than foresight. Looking back to 1910, at the previous appearance of Halley's Comet, we can say confidently that at that time, radio and aircraft were destined to have the most remarkable development and to most affect human society. Now in 1986, with Halley's Comet again in the sky, one can be less sure, as one looks into the cloudy crystal ball of the future.

Nuclear weapons may overtake all our hopes and destroy us. International rivalries may stop short of war yet make it impossible for us to tackle, in a rational manner, such serious problems as overpopulation, pollution, and the erosion of life's amenities, leaving civilization to decline and stumble into wretchedness and misery.

If humanity can avoid catastrophe, however, we may expect to advance chiefly by way of robots and rockets.

Robots offer us a way out of a dilemma that has been dehumanizing humanity since the coming of agriculture. The vast majority of the tasks human beings must undertake have always seriously underutilized the complex and versatile human brain, which in the long run has brought it to atrophy and ruin in most individuals. Yet there has seemed no way out. Although most jobs are too simple for the human brain, they nevertheless remain too difficult for animals or noncomputerized machines.

With the Industrial Revolution, beginning in 1780, there was the introduction of power machinery (steam engines, internal-combustion engines, electric motors) that lifted the heavy weight of meaningless muscular labor from the backs of humanity. Now, for the first time, we have computerized machinery (robots) that can lift the heavy weight of meaningless labor from the minds of humanity.

Naturally, as robots take over, the first effect will be that jobs for human beings will disappear. There will be a difficult transition period in which human beings must be retrained to do other work, or, failing that, be allowed to work at what they can, while a new generation is educated in a new fashion to be suited to more creative work. It might be argued too that few people are naturally creative to make a creative society possible; but it was once widely felt that too few people had a natural gift for reading and writing to make a literate society possible. The suspicion was wrong then, and it is probably wrong now.

Meanwhile, new jobs of many kinds (different and more fulfilling) will come into being, and a great many of them will be involved with space, for rocketry will extend the human range widely.

A space station will be built on which dozens of men and women, in shifts, will build elaborate space structures. Power stations will be built that will trap the energy of the Sun and turn it to radiation that will be beamed to Earth, where it will be converted into electricity. This may solve Earth's energy needs permanently.

Mining stations will be built on the Moon. Lunar ores will be hurled into space where they can be smelted into metals and converted into cement and glass. This will relieve Earth itself of the necessity of supplying all but a small fraction of the materials the space society will need.

Factories will be built in space where the unusual properties of the surroundings (zero gravity, endless vacuum, hard radiation from the Sun, etc.) will make new products, methods, and techniques possible. Increasingly, industrial projects will be transferred from Earth, so that the planet can be freed of some of the disadvantages of industrialization while not losing any of the benefits.

Much of space work can be done by automated procedures and the use of robots, but even so, space settlements will be put into orbit, each one capable of holding ten thousand to a hundred thousand people and each one designed to have a benign environment. Each may have a culture of its own choosing so that humanity will display greater variety and a greater spread of creativity and aesthetics than ever. And it is these settlers, more accustomed to space and to artificial environments, upon whom we can depend for future journeys to Mars, the asteroids, and beyond.

I have already listed what observers on Halley's Comet might have noted on Earth on each of its earlier passings. What, then, might they note on Earth when they next return in 2062?

Perhaps, if robots and rocketry *do* play the role I have projected for them, something like this will be observed:

The cometary observer will be looking at an Earth that has for three-quarters of a century been facing the possibility of a population of eleven billion by 2062. More and more desperate measures have had to be taken to achieve population stability and the population has leveled off at seven billion. While Earth is not yet filled with brotherhood and love, the deadly danger of nuclear war has burned itself into the minds of humanity, and has produced a somewhat resigned acquiescence in international cooperation.

Moreover, the space between the Earth and the Moon is swarming with activity. In the face of an enormous constructive project, clearly designed to relieve the pressures on Earth, international rivalry looks petty and foolish (many say "insane") each year.

Half a dozen space stations are in orbit about the Earth, all internationalized, so that Americans, Russians, Japanese, Europeans, Brazilians, Arabs, Africans, all work together. There is a mining station on the Moon, not yet producing much but with a magnetic driver in place and with an experimental smelting station in orbit about our satellite. One solar power station is in operation. Its energy production is small scale, but it is experimental, and much larger stations are on the drawing board and await the growth of the Lunar mining station.

There is an astronomical observatory in space, and another, larger one on

the far side of the Moon. Two factories exist in space, each thoroughly ani-
mated, one manufacturing electronic components, another developing new al-
loys. Plans for a biochemical laboratory for carrying on the perhaps dangerous
experiments of genetic and biological research are well advanced.

Robots swarm everywhere, on Earth as well as in space. The first large-size
space settlement is almost done and an international commission is trying to
choose settlers from among the millions who are vying for permission to
live there.

The cometary observer has never seen anything like this and marvels over
the great change since his last visit. He suspects that, barring catastrophe, when
the comet next returns in 2138, it will witness scores of space settlements, a
domed city on the lunar surface, and mining stations on a dozen of the
larger asteroids.

And, thereafter? Who can say?

22

ICE IN ORBIT

Far, far away from us, one or two light-years away, as many as one hundred billion icy objects, each only a few miles across, slowly circle our Sun. At their distance, about three thousand times as far away as Pluto, the most distant known planet, the Sun seems to be only a bright star. It would be the brightest star in the sky, but not by very much.

Moving slowly under the weak pull of the distant Sun each of the icy objects takes millions of years to complete one turn about their huge orbit. From the time the solar system was first formed they have circled the Sun several thousand times.

Left to themselves, these objects would circle the Sun forever in unchanging orbits, but they are not left to themselves. The nearest stars also pull at them. Those pulls are strong enough to speed the motion of some comets and slow the motion of others (depending on the position of each relative to a particular star).

When an object's motion is increased in this manner, it is forced to move farther from the Sun. If a particular object happens to be pulled into increased speed a number of times, it may eventually move outward so far as to be lost to the Sun. It would then go wandering off into the empty spaces between the stars.

But suppose an object happens to be pulled by a star in such a way that it is made to move more slowly. That object then drops inward toward the Sun. Or what if two of the objects collide and, as a result of the collision, one of them is brought almost to a dead halt. The one that is halted drops almost directly toward the Sun (just as an airplane would drop toward the ground if it suddenly halted in midair).

Once these objects drop toward the Sun, startling changes begin to take place.

Far out in their original orbits, the temperature of these objects is near absolute zero. They are composed of substances such as water, ammonia, methane, carbon dioxide, and cyanogen, which are built up of the common elements, hydrogen, oxygen, carbon, and nitrogen. All these substances freeze into icelike solids at very low temperatures. Trapped in the ice are bits of dust and tiny gravel made up of rock and metal. Some of the frozen objects may have a small core of rock or metal at the center.

Once such an icy object is slowed and is caused to drop in the direction of the Sun, it begins to warm up. By the time it is only a few hundred million miles from the Sun, the icy solids are evaporating into gas. The dust and gravel are liberated and, hovering above the surface of the object, reflect the sunlight and glitter as a result. Astronomers, observing the object through a telescope, do not see a point of light (as they would if the object were still frozen), but see a small circular cloud of haze instead.

As the object approaches still closer, and warms up still more, the circle of haze expands. This haze is affected by the Sun, for the Sun is always emitting tiny electrically charged particles that go speeding off in all directions. These speeding particles are called the "solar wind."

The solar wind sweeps some of the haze into a filmy "tail" that naturally points away from the Sun. As the object approaches the Sun more and more closely, the tail grows longer, larger, and brighter. If the object comes close enough to the Earth, it can eventually be seen with the unaided eye—a hazy spot of light with a long hazy tail attached. It swings around the Sun and then retreats back to the vast distances from which it came, growing fainter and fainter. It may be hundreds of thousands of years, or even millions of years, before it comes again.

However, if such an object happens to pass near one of the larger planets as it moves toward the Sun, the gravitational pull of the planet may alter its orbit into a much smaller one. The object may then return after only a hundred years or less.

Actually, no astronomer has ever seen this far distant belt of objects and there is no direct evidence for its existence. However, that belt seems the only way of explaining the coming of these strange and beautiful hazy objects with their long tails.

Human beings saw these objects for many centuries before astronomers learned what they were. In the absence of knowledge, these objects were frightening indeed. Everything else in the sky was a circle or a dot of light and behaved in a regular fashion. Early astronomers quickly learned the manner of motion of the stars and planets and could predict such changes as the phases of the Moon, the shifting positions of the various planets against the starry background, eclipses of the Sun, and so on. Since the heavenly bodies were thought to influence human lives, their regular motion was a calming reassurance that events on Earth would proceed regularly and properly.

But then there were these hazy objects, coming out of nowhere, moving across the sky in an unpredictable way, and then vanishing into nowhere. These seemed to be special revelations from the gods, a warning of irregular events.

Surely such events could scarcely be happy ones. Consider the appearance of the object. To some, the tails seemed to be shaped like long, curved swords—the weapon of the angel of death. To others, the tails seemed like the unbounded hair of a woman in mourning as she went shrieking across the sky. In fact, the Greeks called these objects "hairy stars." The Greek word for "hairy" is "kometes"; giving the word its Latin spelling, we have called these tailed objects "comets" ever since.

When a comet appeared, people in Europe were sure it meant disaster—war, famine, plague, death. Panic spread across the land as people turned to prayer or to magical practices to save them from disaster. And, as a matter of fact, in every year in which a prominent comet appeared, there was always disaster. Some famous man would die that year or a few years before or a few years after. There would be a famine somewhere and a plague somewhere, and wars would break out, or had already broken out. This was considered absolute proof that comets were fearful and dangerous.

Of course, in those years in which comets did *not* appear, similar disasters (sometimes worse ones) also took place, but somehow people took no notice of that.

In 1687, however, the English scientist Isaac Newton worked out the "law of universal gravitation," which could be used to describe the way in which the Moon moved around the Earth, and the way in which the Earth (and the other planets) moved around the Sun.

Newton's young friend, Edmund Halley (the name rhymes with "valley"), who had put up the money to publish the book in which Newton described his law, wondered whether it could be used to describe the motion of comets, too. There had been a bright comet in the sky in 1682 and Halley had taken careful observations of it, noting the exact path it had taken across the sky.

He also studied observations of previous comets. He noted, for instance, that there had been a comet in 1607 that had been carefully observed and that had taken the same path across the sky that the comet of 1682 had. Going further back in time, comets that had appeared in 1531 and 1456 had taken that path also.

These comets had appeared at intervals of seventy-five or seventy-six years. It struck Halley that all four appearances were of the same comet. He worked out an orbit that would account for this and it turned out to be a long cigar-shaped one, with the Sun (and Earth) near one end. When the comet was near that end, it reflected sunlight, formed a tail, and was visible. When it retired toward the other end, it dimmed and was no longer visible, but it was still there.

Halley predicted that the comet would return in 1758. He could, however, scarcely expect to see it, for in that year he would be 102. Actually, he lived to be eighty-five and died in 1742, sixteen years too soon to see whether his prediction would come true or not.

Most astronomers didn't take Halley's prediction seriously, and no important search was set up when 1758 rolled around. However, a well-to-do German farmer named Johann Georg Palitzsch whose hobby was astronomy set up a telescope in his fields, and every night, if the sky were clear, he carefully studied the spot where the comet ought to appear.

The year had almost gone and Palitzsch had seen nothing, but then, on Christmas night, December 25—there it was. The comet had returned on schedule, and ever since it has been known as "Halley's Comet" or, in recent years, "Comet Halley." Because it was the first comet to be shown to be an ordinary member of the solar system, and because of all the comets that return every one hundred years or less ("short-period comets") it is by far the brightest, it became, and still is, the most famous and watched-for of all comets.

Going backward in time, over thirty appearances of the comet can be found in the history books. It appeared at the time of the Norman conquest of England, for instance, at about the time of the birth of Jesus, and so on.

After 1758 Halley's Comet appeared again in 1835 and in 1910. The great American writer, Mark Twain, was born in 1835 when Halley's Comet was in the sky, and in 1910, when it was again in the sky, he lay dying. When friends tried to assure him he would recover, he said, "No, I came with the comet and I shall go with the comet," and he did.

After Halley's Comet's return in 1758, there was a period of time during which astronomers went mad over comets. Almost every astronomer tried to discover a few that would be named for himself. The champion in this respect was a French astronomer, Charles Messier. He spent fifty years at his telescope and discovered twenty-one comets. Every time someone else discovered a comet he was riddled with jealousy. When his beloved wife died, he mourned her sadly, but he had to admit to a friend that all the time he watched at her bedside he kept worrying that he might miss discovering some comet.

Every time a comet passes the sun, it loses a quantity of its ice, which vaporizes and never returns. Eventually, the loss will break a comet into two smaller pieces or cause it to vanish altogether, leaving only a trail of dust and gravel over its orbit. Astronomers have seen comets break up and vanish.

Halley's Comet will be no exception. It is undoubtedly much smaller than it was when William of Normandy invaded England, and it becomes smaller at each return. Some day it will no longer be visible to the unaided eye, and eventually it might not return at all.

When it returns in 1986, it will not be much of a sight. Earth will be in a part of its orbit that will place it rather far from the comet; so the comet won't appear very bright. When the comet is at its brightest, the tail will be pointing toward us—so we won't see that most spectacular feature broadside. Finally, the position of Halley's Comet will be such that it will be best seen from the Southern Hemisphere. (There are people who are planning to go to Buenos Aires or Capetown to see the comet.)

It is a shame, in a way, for in the 1800s there were no fewer than five giant comets in addition to Halley's that came in from the far-off comet shell. They developed tails that stretched half way across the sky, and then they moved back to the shell; they won't be back for a million years or so. Since 1882, not one giant comet has appeared, and to see Halley's Comet again, we will have to wait till 2058, if any of us live that long.

Some comets skim fairly close to the Earth as they pass around the Sun. Is it possible that one might strike the Earth sometime? That *is* possible and such a strike might be very damaging. In 1908 there was a gigantic explosion in northern Siberia which flattened every tree for forty miles about and wiped out a herd of deer. But it (fortunately) did not kill a single human being, for that section of the land was a total wilderness. It was thought a meteorite had struck, but such an object, made of metal or stone, would have left a crater, and no crater was found. Nowadays, it is thought that a small comet struck the Earth in 1908, evaporating before it reached the ground, but in the process producing a devastating explosion.

Many scientists now believe that a much worse cometary strike took place sixty-five million years ago. They believe it kicked so much dust into the stratosphere that sunlight was totally blocked for months, or perhaps years, so that most of life died on Earth, including all the magnificent dinosaurs. It may even be that some such strikes take place every twenty-six million years or so, since life seems to undergo a process of "Great Dyings" at these intervals.

One suggestion is that the Sun has a very dim companion star which circles it at a huge distance. At one end of its orbit it comes close enough to plow its way through the innermost comet shell where the comets are thickest. The star's gravitational pull would cause millions of comets to drop down toward the Sun and a few of them would be sure to hit the Earth.

But don't worry about it just yet. This suggestion is a very chancy one and may not be true, and, even if it is, that companion star is at its greatest distance from us right now and won't be doing any damage for another thirteen million years or so.

Afterword: The preceding essay was a second response to Halley Hysteria. In all my responses I steadfastly refused to discuss the question of where and how to watch for the comet. I even refused to do it in my book Asimov's Guide to Halley's Comet, *published by Walker and Company in early 1986. My reason for doing so was simple. It was going to be a very poor show and everyone would be disappointed, and I wasn't going to make believe there was going to be anything to see.*

The result was that critics objected to my book for failing to include that useless information. What's more, the preceding essay was rejected by the first magazine to which I sent it when I refused to add a section on viewing the comet. However, I sold it on my second try, and I think I was right in clinging to my principles.

23

LOOKING FOR OUR
NEIGHBORS

Here we are, living on a middle-sized planet, circling a middle-sized star. That star is part of a galaxy which includes a couple of hundred billion other stars, and beyond our galaxy are a hundred billion other galaxies, each with anywhere from a few million to a few trillion stars.

Altogether there are about 10,000,000,000,000,000,000,000 stars in the Universe. We intelligent beings, *Homo sapiens,* are on a planet circling one of them. Can we honestly think that nowhere among all those other stars is there another planet carrying intelligent beings? Can we be alone in so large a Universe?

And if we're not alone, if we have neighbors, might it not be that some of them are trying to signal us? If so, let's consider—

1. *How would they try to signal us?*

It would have to be with some sort of signal that can cross the vast emptiness of interstellar space. There are, indeed, objects that reach us across interstellar space. There are, for instance, charged particles such as those that make up cosmic rays.

Charged particles, however, aren't suitable for the purpose. Their paths curve in magnetic fields, and every star has a magnetic field. Our galaxy has an overall magnetic field. Charged particles therefore come looping in on the bias, and though they end by reaching us from some particular direction, we haven't the faintest idea from that what their *original* direction of travel was, so that we can't tell where they came from. Such signals would be useless.

Uncharged particles travel in a straight line regardless of magnetic fields. If they are without mass, they travel through vacuum at the speed of light, the maximum velocity. There are three kinds of massless, uncharged particles: neutrinos, gravitons, and photons. Neutrinos, which are liberated by the nuclear reactions going on within stars, are almost impossible to detect. Gravitons, which are associated with gravitational fields, are even harder to detect. That leaves photons, which are easy to detect.

Photons are the particles of electromagnetic radiation, all of which are made up of waves. There are two types of photons that can make their way easily through our atmosphere. There are the photons associated with waves of visible light, and the photons associated with microwaves, the waves of which are about a million times longer than those of light.

The signals would be coming from a planet that is circling a star. Every star sends out a great deal of light. If intelligent beings on a planet send out a light signal, that signal might be drowned out by the starlight. On the other hand, ordinary stars are not very rich in microwaves and a microwave signal would stand out clearly. And if the proper instruments are used, microwaves are even easier to detect than light is.

Microwaves come in many wavelengths. Which wavelength should we watch for?

During World War II, a Dutch astronomer calculated that cold hydrogen atoms in deep space sometimes undergo a spontaneous change in configuration that results in the emission of a microwave photon that is twenty-one centimeters in wavelength. Individual hydrogen atoms undergo the change very rarely but if there are a great many hydrogen atoms involved, great numbers of photons would be emitted every moment and these could be detected. In 1951 an American physicist *did* detect them, for "empty" space contains a thin scattering of hydrogen atoms that mount up if you consider cubic light-years of volume.

This twenty-one-centimeter wavelength is everywhere so it is of prime importance in studying the properties of the Universe. Any intelligent species would have radio telescopes designed to receive such signals, and that wavelength would therefore be a natural for deliberate signalling.

If our astronomers ever got a beam of twenty-one-centimeter radiation that contained hardly any other wavelengths, they would become suspicious. If the radiation went on and off, or got stronger and weaker, in a manner that was not entirely regular, and not completely random either, then they would know someone was trying to tell us something.

2. *But where should we listen?*

It would be exceedingly expensive and time-consuming to try to listen to every star, so we should start with those stars that seem the most likely signallers, stars that have a reasonable chance of possessing a planet inhabited by intelligent beings.

There might be all kinds of life in the Universe, but our own form of life is built on the most common elements there are: hydrogen, oxygen, carbon, and nitrogen. It has a water background and a backbone of complex carbon compounds. We have, as yet, no evidence that life can exist on a different chemical basis, so we should assume that life elsewhere is fundamentally like our own.

Stars that are considerably more massive than our Sun are also considerably brighter and must consume their hydrogen fuel very rapidly to keep from collapsing. Their lifetime is considerably shorter than that of our Sun, and probably not long enough to allow the slow processes of evolution needed to develop a highly intelligent species, if life there is fundamentally like our own.

Stars that are considerably less massive than our Sun are so dim that a planet would have to circle it at a very close distance to be warm enough to possess liquid water on its surface. At such close distances, tidal influences would slow the planetary rotation and produce temperature extremes during the long day and long night that would not be suitable for life that is fundamentally like our own.

It makes sense, therefore, to concentrate on those stars that are between 0.8 and 1.2 times the mass of the Sun.

At least half the stars in the Universe are part of binary systems. It is possible for one or both stars of such a binary system to have planets in stable orbits, but there's less chance of it than in the case of a single star like our Sun. Therefore, we ought to concentrate on Sun-like single stars.

Naturally, the closer a star is, the less likely it is for its signal to fade with distance; so the stronger the signal, the more likely we are to detect it. Therefore we ought to concentrate, to begin with, on the closest Sun-like single stars.

Some of them are sure to be far in the southern skies and invisible from northern latitudes; or, if visible, they would always be near the southern horizon. Since the best astronomical equipment we have is concentrated in the northern hemisphere, it makes sense to concentrate on the closest Sun-like single stars in the northern sky.

There are three of these, all about 0.8 times the mass of the Sun. They are Epsilon Eridani (10.8 light-years away), Tau Ceti (12.2 light-years away), and Sigma Draconis (18.2 light-years away).

In 1960 the first real attempt was made to listen to the twenty-one-centimeter wavelength. (The attempt was called "Project Ozma.") The listening began on April 8, 1960, with absolutely no publicity since the astronomers feared ridicule. It continued for a total of 150 hours through July, and the project then came to an end. The listening concentrated on Epsilon Eridani and Tau Ceti, but nothing was heard. The search was very brief and not very intense.

Since Project Ozma, there have been six or eight other such programs, all at a level that was more modest still, in the United States, in Canada, and in the Soviet Union. No signals were picked up.

In 1971 a group at NASA began thinking about something called "Project Cyclops." This would be an array of 1,026 radio telescopes, each one hundred meters in diameter. They would be placed in rank and file, and all of them would be steered in unison by a computerized electronic system.

The array would be capable of detecting, from a distance of a hundred light-years, the weak radio waves that leak out of equipment on Earth. A deliberately emitted message signal from another civilization could surely be detected at a distance of at least a thousand light-years. This could make it possible to listen to a million different Sun-like single stars, not just two or three very close ones.

Such an array of radio telescopes would cost anywhere from ten to fifty billion dollars—but remember that the world spends five hundred billion dollars on armaments every single year.

3. *But why should we bother?*

Well, suppose we do detect signals from some other civilization. Undoubtedly, they will be more advanced than we will be, and if we can interpret what it is they are saying, we may discover a great deal about the Universe that we don't yet know and might not be able to find out for many years on our own. We would get a free education; or if not quite free, a priceless one.

Even if we learned *one* new thing, something that didn't seem very important in itself, it might give us a head start in a new direction. It would be like looking up one key word in the back of a book of crossword puzzles. It could give us the one clue we need to work out a whole group of words.

Even if we didn't learn anything, because we found we couldn't decipher the signals, the effort of deciphering might itself teach us something about communication and help us with our psychological insights here on Earth.

And even if there was no chance at all of deciphering the message and if we didn't even try, there would be important value just in knowing for certain that out there on a planet circling a certain star was another intelligent species.

It would mean that we were not alone in the Universe, and it might force us to take a new look at the world and ourselves. We have been so used to thinking of ourselves as lords of the world and the crown of creation that we have been acting in a dangerously arrogant way. It might do us good to start thinking of ourselves as one of many, and by no means the greatest. For one thing, it might start us thinking of ourselves as Earthmen, and it might encourage us to cooperate.

It would mean that it was possible for at least one other intelligent species to develop a technology more advanced than our own and to survive. At least one other species would have survived nuclear weapons, overpopulation, pollution, and resource depletion. If they could do it, maybe we could, too. It would be a healthy antidote to despair.

Finally, even if we found nothing at all, *nothing,* it would still be worth it.

The very attempt to construct the necessary equipment for Project Cyclops would surely succeed in teaching us a great deal about radio telescopy and would undoubtedly advance the state of the art.

If we searched the sky with new equipment, new expertise, new delicacy, new persistence, new power, we would surely discover a great many new things about the Universe that have nothing to do with advanced civilizations and that don't depend on detecting signals. We can't say what those discoveries would be, or in what way they would enlighten us, or just what direction they could lead us. However, knowledge, *wisely used,* has always been helpful to us in the past and will surely always be helpful to us in the future.

There is every reason, then, to think that the search for extraterrestrial civilizations would be a case of money well-spent, however much it would cost.

24

LIFE IS WHEREVER IT LANDS

About 4.6 billion years ago, the Sun and its family of planets—including Earth—formed out of a vast primordial cloud of gas and dust.

During the first half-billion years of its existence, Earth underwent the final steps of its formation, gathering the bits and pieces of matter, from pebbles to mountains, that still cluttered its orbit. All of its structure undoubtedly remained in a state of flux during this period, being kept hot and even molten by the energy of violent collision.

In May 1983, geologists found bits of rock in western Australia that seem to be 4.1 billion years old, so the birth pangs may have been over by then. The planetary crust had solidified and some of it, at least, has remained solid through all that time.

It was perhaps 4.0 billion years ago that the crust was cool enough to allow a liquid ocean to accumulate, and only after that was life (as we know it) possible. In fact, the earliest bits of microscopic life have left their exceedingly tenuous traces in rocks that are perhaps 3.5 billion years old.

It would seem, then, that within half a billion years after Earth more or less achieved its present form, life already existed in its seas.

That means that proteins and nucleic acids of considerable complexity must have formed by that time. And unless we are willing to suppose that these were produced by divine fiat, they must have been preceded by a long period of "chemical evolution." In other words, simple molecules of the type we suppose to have existed on the primordial Earth must slowly have formed more and more complex molecules through processes in accord with the laws of physics and chemistry. Eventually, molecules were formed that were sufficiently complex to display the fundamental properties of life.

We can make shrewd guesses, but we don't know for certain what the physical and chemical properties of Earth's crust, ocean, and atmosphere were before life appeared. We also are not certain as to the amount and nature of the forms of energy that bathed and permeated the terrestrial environment of the time. We can experiment with sterile mixtures of chosen substances exposed to selected forms of energy in such a way as to form a system that we believe may duplicate the prelife environment, and then draw deductions from the results, but these are bound to remain uncompelling.

We are faced, in short, with having to explain the formation of something fearfully complex in a surprisingly short time under conditions that we can only vaguely guess at.

How it would help if we could find some nearby world that we could study in detail, a world on which life did not exist but on which chemical evolution was already under way. Even if conditions were so unfavorable that life could not conceivably develop, and chemical evolution was sure to be aborted and brought to a halt in its early stages, the hints we would receive would extend our knowledge far beyond our present near-zero level, and would help us enormously in our attempts to sketch out the steps by which life came to be.

There were hopes for the Moon, but they failed. There were hopes for Mars, but they failed, too. On neither world was there any sign of organic material in the soil. There are now some flickers of hope that there may be signs of chemical evolution on some worlds of the outer solar system, such as Europa and Titan, but they are not very strong.

And yet, oddly enough, even though the worlds of our own solar system have failed us, and even though we can examine no other planetary systems, all is not lost. Chemical evolution exists elsewhere than on Earth, and it has been observed.

The Universe as we know it came into existence perhaps fifteen billion years ago, and our galaxy was ten billion years old before our Sun and its planets formed. All that time, the cloud of dust and gas out of which the solar system formed remained in existence, more or less undisturbed, until something occurred to trigger its condensation. It is not surprising that elsewhere in the Galaxy (and in other galaxies, too of course) such interstellar clouds still exist, *still* undisturbed over an additional five billion years.

In fact, there are thousands of such clouds, each many light-years in diameter, and each with masses dozens of times as large as our Sun.

What do such clouds consist of?—Obviously of the kind of atoms that go to make up stars and planetary systems, since that is what they form on condensation. The most common atoms to be found there, as we can tell by spectroscopic studies, are those of simple atoms with particularly stable nuclei—hydrogen, helium, carbon, nitrogen, oxygen, and neon.

Of these six major atomic components, helium and neon are inert and their atoms remain isolated and unaffected by any conditions less extreme than those existing deep within a star. The remaining four, however, hydrogen, carbon, oxygen, and nitrogen, are precisely those out of which the bulk of living tissue is formed.

Interstellar clouds are "clouds" only in comparison to the nearly empty space outside them. Matter in such clouds is very rarefied compared, let us say, to Earth's atmosphere; so rarefied, in fact, that astronomers would not have been surprised if atoms existed within them in the single state only. After all, in order for atoms to combine, they must first collide, and collisions would be expected to be rare indeed in matter as rarefied as it is in interstellar clouds.

Atoms and molecules can emit or absorb electromagnetic radiation characteristic of themselves, and by observing such emission or absorption astronomers can identify, with an approach to certainty, atom combinations present in interstellar clouds. In 1937, observing in the visible light wavelength region, astronomers detected carbon-hydrogen and carbon-nitrogen combinations (CH and CN) in the clouds.

It was not thought that much more was ever likely to be found, but after World War II radio astronomy came into being. It then became possible to detect, with increasing precision, specific wavelengths in the microwave region. Microwaves have much longer wavelengths than light waves do, and are much less energetic in consequence. This means that microwaves can be emitted even by quite cool material, and this makes it possible to locate and identify relatively small concentrations of other types of atom combinations.

In 1968 astronomers detected the first atom combinations in outer space that were made up of more than two atoms. These were water, the molecules of which are made up of two hydrogen atoms and an oxygen atom (H_2O), and ammonia, the molecules of which are made up of three hydrogen atoms and a nitrogen atom (H_3N). With that, the new science of "astrochemistry" was born.

In the first fifteen years of astrochemistry, astronomers have detected an astonishing variety of atom combinations in interstellar clouds, dozens of them. As techniques have improved, the combinations discovered have been more and more complex, and almost all but the simplest have involved the carbon atom.

By 1971 the seven-atom combination of methylacetylene was detected—a molecule made up of four hydrogen atoms and three carbon atoms (CH_3CCH). In 1982 a thirteen-atom combination was detected, cyano-decapenta-yne, which consists of a chain of eleven carbon atoms in a row, with a hydrogen atom at one end and a nitrogen atom at the other ($HC_{11}N$).

We don't know what the chemical processes are that give rise to these multi-atom combinations, any more than we know what the chemical processes were that gave rise to complex chemicals in Earth's crust during the prelife stage. The difference is that in the interstellar clouds, we can observe the results. We may not be certain how CH_3CCH and $HC_{11}N$ are formed, but we know those combinations are there. That lowers the degree to which we must guess by a notch or two.

So the one place outside Earth that we know by observation that chemical evolution is taking place is in the interstellar clouds.

To be sure, conditions in the interstellar clouds are enormously different from those on Earth's surface, so we can't reasonably suppose that we can easily apply the lessons learned from one to the problems posed by the other. Nevertheless, we can say this much: In the course of billions of years, the chemical evolution taking place in interstellar clouds is producing complex atom combinations built about a carbon-atom skeleton, and that is just what we find in living tissue, too.

The carbon combinations in interstellar clouds are, to be sure, far, far less complicated than those in living tissue. However, it seems reasonable to suppose that the more complicated a combination is, the longer it takes for the very occasional atom collisions in interstellar clouds to build it up, and the smaller its concentration. The smaller its concentration, the feebler its total microwave emission, and the less likely we are to detect it.

It is not impossible, however, that various amino acids may have been built up in interstellar clouds. (It is the amino acids that are the building blocks of protein molecules—one of the two types of giant molecules present in all life.) If they are there, they may well be present in such minute concentrations that they cannot possibly be detected—and yet over the entire volume of the cloud they may be present in millions of tons.

Will this remain idle speculation forever, or is there any possibility of gaining some observational data that will lend support to the view?

Well, the cloud out of which the solar system formed had been undergoing chemical evolution for ten billion years and must have had a great many atom combinations present within itself, perhaps including amino acids and molecules of even greater complexity. As it condensed to form the Sun and the planets, however, the great heat that developed, as the Sun ignited its central nuclear fusion and the planets were heated to melting by endless collisions, would surely have destroyed all such combinations and left no trace of them.

The solar system, however, does not consist of the Sun and its planets only. It also includes myriads of small, even tiny, bodies. The smaller an object, the less likely it is to have been heated to melting in the process of its formation, and the more likely it is to represent matter as it was at the time the cloud existed. It may be, then, that in the very smallest asteroids, we may find traces of compounds originally formed by chemical evolution in the cloud.

We might conceivably be able to check this since some very small objects are continually colliding with Earth. The smallest are heated to vapor by the resistance of Earth's atmosphere. The larger ones survive, in part at least, and strike the Earth's surface as "meteorites."

Most meteorites are stony in nature, and a few are metallic, and, in either case, they do not possess a perceptible quantity of carbon. A very few meteorites, the "carbonaceous chondrites," *do* contain carbon. One such fell near Murray, Kentucky, in 1950, and another exploded over the town of Murchison in Australia in 1969.

These two have been studied extensively, the carbon-containing fraction being extracted and analyzed. Among the molecules present in the meteorites were amino acids, six of them of types common in Earthly proteins, the rest not too distantly related.

As it happens, amino acids fall into two types, "L" and "D." When amino acids link up in chains to form proteins, the chains are either all-L or all-D. A chain made up of both varieties doesn't fit together well and would easily break up even if it formed in the first place.

On Earth the protein chains are all-L, though that may be just chance, with an all-L chain happening to be the first to be complex enough to show properties of life, so that it reproduced itself and took over the Earth before an all-D molecule had a chance to form at the proper level of complexity. As a result, living organisms on Earth form only L-amino acids, and D-amino acids are very rare in nature.

When chemists form amino acids in the laboratory by methods that don't involve living tissue, L and D varieties form in equal amounts. The amino acids in the carbonaceous chondrites are also found as L and D in equal amounts, thus showing they were formed by chemical processes not involving life.

In 1982, however, one of the fragments of the Murchison meteorite was reported to contain amino acids that were predominantly L. It may be that it was contaminated by Earthly material, or that there was something wrong with the experimental procedure—but it might also be that there was, or had been, something more complicated than amino acids present at one time. Might there have originally been a bit of all-L protein present that, at some point, broke up into L-amino acids only? We can't tell.

Will we ever be able to tell?

Perhaps! But not through the study of asteroids and meteorites. The asteroid belt is relatively close to the Sun and the Sun's energetic radiation (ultraviolet and beyond), unfiltered by atmosphere, would tend to break up the carbon-atom combinations. This is especially true of meteorites, which must approach the Sun even more closely than ordinary asteroids do or they wouldn't be in a position to collide with Earth.

However, there are also tiny bodies at the very edge of the solar system, far beyond the orbit of Pluto. It is thought that at a distance of one or two light-years from the Sun, there is a vast cloud of comets that enclose the solar system like a sphere. These may be the remnants of the outermost region of the cloud out of which the solar system formed. They are thought to consist of icy materials built up of hydrogen, carbon, nitrogen, and oxygen atoms mainly, and since they are so far from the Sun, no substantial quantity of energetic radiation has struck them. They have circled the Sun, undisturbed for billions of years, and may still contain all the products of chemical evolution.

We can't, as yet, possess much hope of examining the comet cloud by astronaut or probe, but some comets, for one reason or another, occasionally fall out of the cloud and approach the inner solar system. Once that happens, such a comet takes up a new orbit that brings it into the inner solar system periodically. With each pass through our part of the system, it is exposed to heat and radiation from the Sun which disturbs and eventually destroys it. Still, if one that is approaching for the first time (as Comet Kahoutek was thought to have done in 1973) it could actually be studied at close range by astronauts, and it might be able to tell us how far chemical evolution proceeded in the cloud out of which we formed.

In fact, the comets may be the key to the formation of life.

It may be that chemical evolution on Earth need not be viewed as having originated from scratch. It may be that every once in a while a comet or a fragment of one collides with a planet under circumstances that do not quite destroy all the molecules of which the comet is composed. It may be that if a planet is, like Earth, potentially hospitable to life, the molecules will then accumulate and continue their chemical evolution.

With such a head start, it is less surprising that life would form in as little as half a million years. What's more, a similar process might take place in every planetary system, and, if so, that makes it all the more possible that life would develop on any planet that had the capacity to be hospitable to these comet-inspired beginnings.

(P.S. Fred Hoyle and Chandra Wickramasinghe speculate that chemical evolution in interstellar clouds and, therefore, in comets, actually reach the stage of living viruses, so that planets might be seeded with life directly. They also suggest that approaching comets may infect a planet such as Earth with new strains of pathogenic viruses that explain our periodic plagues. These are extreme views, however, that astronomers generally are not willing to take seriously.)

25

EINSTEIN'S THEORY
OF RELATIVITY

According to the laws of motion first worked out in detail by Isaac Newton in the 1680s, different motions add together according to the rules of simple arithmetic. Suppose a train is moving away from you at twenty kilometers per hour and a person on the train throws a ball at twenty kilometers per hour in the direction of the train's motion. To the person moving with the train, the ball is moving twenty kilometers per hour away from him. To you, however, the motion of the train and the ball add together, and the ball is moving away from you at the rate of forty kilometers per hour.

Thus, you see, you cannot speak of the ball's speed all by itself. What counts is its speed *relative* to a particular observer. Any theory of motion that attempts to explain the way velocities and related phenomena seem to vary from observer to observer would be a "theory of relativity."

Albert Einstein's particular theory of relativity arose from the fact that what works for thrown balls on trains doesn't seem to work for light. A moving source of light might cast a beam of light in the direction of its travel, or against the direction of its travel. In the former case, the light should (according to Newton's laws) travel at its own speed *plus* the speed of the light source. In the latter case, it should travel at its own speed *minus* the speed of the light source. That's the way the thrown ball on the moving train would behave.

But light doesn't behave so. It always seems to travel at its own speed, regardless of the motion of the source. Einstein attempted to describe the laws of the universe to account for that.

Einstein showed that, in order to account for the constancy of light's velocity, one had to accept a great many unexpected phenomena. He showed that objects would have to grow shorter in the direction of their motion (shorter and shorter as their motion was faster and faster, until they would have zero length at the speed of light), that moving objects would have more and more mass the faster they moved, until their mass was infinite at the speed of light, and that the rate at which time progressed on a moving body decreased the faster it moved, until time stopped altogether at the speed of light. For each of these reasons the speed of light in a vacuum was the maximum speed that could be measured.

Furthermore, he showed that a little bit of mass was equal to a great deal of energy, according to his famous equation, $e = mc^2$, where c stands for the speed of light.

All this he worked out in 1905 for bodies moving at the same speed in an unchanging direction. This special kind of motion was dealt with in Einstein's "special theory of relativity."

The effects predicted by Einstein are noticeable only at great speeds. The simple arithmetic of Newton's laws therefore works at ordinary speeds, simply because Einstein's effects are too small to be noticed. Because we are always surrounded by the working of Newton's laws, they come to seem like "common sense" to us, while Einstein's effects seem strange.

Nevertheless, at the time Einstein published his theory, scientists were working with subatomic particles hurled outward by exploding atoms. These subatomic particles moved at great speeds and the Einstein effects could be noted in them exactly. Atom-smashing machines and nuclear bombs couldn't exist unless the special theory of relativity was correct.

In 1916 Einstein worked out a more complicated version of the theory of relativity, one that included not only motion at constant speed in the same direction—but which included any kind of motion at all, motion that changed speed or direction or both. Since it dealt with any motion in general, this version of his theory is known as "the general theory of relativity."

The most common reason why objects move at varying speeds and in changing direction is because a gravitational force is acting on them. A ball when dropped moves faster and faster because Earth's gravity is pulling it. A thrown ball follows a curved path because Earth's gravity is pulling it. The Earth follows a curved path in space because the Sun's gravity is pulling it.

For this reason, Einstein's general theory had to deal with gravitation.

Einstein's equations showed that if there were no matter anywhere, and no gravitation, a moving body would follow a straight line. If there was matter, the surrounding space would be distorted so that the moving body would follow a curved line. Einstein's general theory showed what those curves ought to look like and they weren't quite what Newton's theory of gravitation predicted.

The difference between Einstein's and Newton's equations is very slight indeed, and only the most careful and delicate measurements can show us which set of equations is actually followed by moving bodies.

One way of telling whether Einstein or Newton is right is by studying the behavior of light when it passes through a strong gravitational field. According to Einstein, the light would pass through the distorted space and follow a very slight curved path. This would not be so according to the Newtonian rules.

In 1919 a total eclipse was observed, and the position of the stars near the Sun was accurately measured. If the light travelled a curved path, each star should seem to be a little farther from the Sun than it ought to be. The amount of displacement would depend on how near the Sun the light had passed. Einstein's theory predicted exactly how large the displacement ought to be. The measurements at that 1919 eclipse, and at every total eclipse since then, seem to place the stars where Einstein predicted they would be.

These were hard measurements to make, though, and difficult to get exact. In our space age, we have been able to do better.

In 1969, for instance, two probes were sent toward Mars. When the probes were on the other side of the Sun, the radio waves that they beamed back to Earth had to pass near the Sun and therefore followed a distorted path. Because of the distortion, the path was longer than a straight line would have been, and it took a bit more time for the waves to reach us. It took them a ten-thousandth of a second longer, actually, something which could be measured and which was just what general relativity predicted.

Another prediction of Einstein's general theory is that in space distorted by gravity time slows down.

A large object giving off light emits that light in waves that are longer than they would be otherwise, because the slowing of time in the object's gravitational field keeps one wave going longer before the next one comes out. Even the Sun, though, has a gravitational field that is not strong enough to show a measurable effect.

Just about the time Einstein advanced his theory of general relativity, however, a new kind of star was discovered. This was the "white dwarf." A white dwarf might have all the mass of the Sun squeezed down into a volume no larger than that of the Earth. Its gravitational field at its surface would be two hundred thousand times as intense as the Sun's.

Particular light waves emitted by a white dwarf would have to be longer by a particular amount than those same light waves emitted by the Sun. In 1925 measurements were taken of the light coming from a white dwarf and were found to be longer by just the right amount predicted by Einstein.

It was, of course, very difficult to take the measurement of the light from a white dwarf. Such stars are very dim, and trying to analyze their feeble beams of light is a tricky matter.

It has been found, however, that under certain conditions, a collection of atoms in a crystal would emit gamma rays. Gamma rays are made up of waves like those of light, but much shorter. The crystal emitted gamma rays that had all the same wavelength exactly, and the wavelength could be measured with great accuracy.

If Einstein's general theory were correct, however, the gamma ray wavelength could be slightly changed if the gravitational pull upon the crystal were increased or decreased.

Here on Earth gravitation increases as one gets closer to Earth's center. If you live in an apartment house, Earth's gravity gets stronger with each floor you descend—but so slightly that you can't possibly tell the difference.

But it shows up in the gamma rays. As the crystal was taken down from the attic to the basement, the wave grew very slightly (but detectably) longer, and by exactly the amount predicted by Einstein's theory.

Still another consensus of Einstein's treatment of gravitation in his general theory of relativity is that gravitation, like light, can be given off in waves. In this way, gravitation, like light, can carry energy away from the object giving it off.

Such gravitational waves are given off when an object moves round and round in response to gravitation, or back and forth, or pulses in and out. The Earth, when it moves around the Sun, gives off gravitational waves, according to Einstein. It is losing energy so that it is slowly dropping in closer to the Sun.

Gravitational waves are so weak, however, and carry so little energy that the waves given off by Earth are far too feeble to be detected. The energy they carry off is so tiny over the next few billion years Earth won't drop closer to the Sun by more than a few meters.

It may be, though, that somewhere in the Universe there are more intense gravitational fields than there are here in the Solar System. Extremely rapid motions may involve the giving off of gravitational waves that are just barely energetic enough to detect.

Many scientists have been trying to detect such gravitational waves using very delicate instruments. Some have reported success, but the results were very borderline, and such reports have not yet been accepted.

In 1967, however, a new kind of star was discovered, a "pulsar." Such a star is even smaller and more compressed than a white dwarf is. A pulsar may have a mass equal to our Sun and yet have it all squeezed into a little ball no more than twelve kilometers across. The gravitational field at the surface of such a pulsar would be 10,000,000,000 times that of the Sun.

In 1974 two pulsars were found that circled each other. With circlings in so incredibly powerful a gravitational field, effects that are barely noticeable in our Solar System become easy to measure.

For instance, there is a small component of Mercury's motion about the Sun that isn't accounted for by Newton's theory of gravitation. A particular point in Mercury's orbit makes a complete circle about the Sun in a little over 3,000,000 years, although by Newton's theory it shouldn't move at all. Einstein's theory, however, accounts for the motion exactly.

That same point in the orbits of that pair of circling pulsars should make a complete circle in a mere eighty-four years, according to Einstein's theory, and the point is moving at the proper speed to do so.

What's more, the pulsars should be giving off gravitational waves as they revolve, much stronger ones than Earth does. The energy carried off by the gravitational waves ought to be enough to cause the pulsars to be spiralling in toward each other at a noticeable rate. This in turn should cause certain radio wave pulses they give off to reach us at shorter and shorter intervals.

In 1978 delicate measurements of those radio pulses were made, and they were found to be coming at shorter and shorter intervals, by just the predicted amounts.

What it comes to is this: In the three-quarters of a century since Einstein advanced his theory of relativity, scientists have tested it in every way they could think of, over and over again, and Einstein's theory has passed every single test without exception.

Afterword: This article was written in February 1979 to help celebrate the centenary of Einstein's birth. Since then, all further tests of general relativity have continued to confirm it strongly. There is now virtually no doubt that Einstein's formulation in 1916 is the best and truest description of the Universe as a whole that we have and, just possibly, that we may ever have.

26

WHAT IS THE
UNIVERSE MADE OF?

What is the Universe made of? All the countless myriads of things, living and non-living, large and small, here and in the farthest galaxies, can't really be countless myriads. That would be too complex, too messy to suit our intuition—which is that the Universe is basically simple, and that all we need is to be subtle enough to penetrate that simplicity.

The Greeks suggested the Universe was made up of a few "elements," and some supposed that each element was made up of invisibly small "atoms" (from a Greek word meaning "indivisible") which, as the name implied, could not be divided into anything smaller.

Nineteenth-century chemists agreed in essence. But what nineteenth-century chemists found were elements by the dozens, each with its characteristic atoms. Again, too complex and too messy.

In 1869 Dimitri Mendeléev arranged the elements into an orderly "periodic table," and in the early twentieth century the rationale of that periodic table was worked out.

It seemed that the atoms were not indivisible after all. They were made up of still smaller "subatomic particles." Each atom contained a tiny nucleus at the center, and this was in turn composed of comparatively massive "protons" and "neutrons," the former with a positive electric charge and the latter uncharged. Outside the nucleus were "electrons"—much less massive than protons or neutrons—which carried a negative electric charge.

By altering the numbers of protons, neutrons, and electrons, the nature and properties of every different type of atom could be explained, and scientists could claim that out of these atoms everything in the Universe was built up. For a while, in the 1930s and 1940s, it seemed that the ultimate constitution of the Universe, the ultimate particles, had been deciphered, and the result was satisfactorily simple. Three different types of particles made up everything.

But there were some puzzles. The electrons were bound to the central nuclei by an "electromagnetic interaction." The negatively charged electron and the positively charged nucleus attracted each other.

Within the nucleus, however, there was no electromagnetic attraction between positively charged protons and uncharged neutrons, and there was a strong electromagnetic *repulsion* between protons. So what held the nuclei together?

In 1935 Hideki Yukawa suggested the existence of what has come to be known as "strong interaction," an attraction among protons and neutrons that is much stronger than any electromagnetic repulsion that existed, and one that decreased in intensity with distance so rapidly that it only made itself felt over subatomic distances.

This explained many subatomic events, but it didn't explain the way in which a free neutron spontaneously changed into a proton, liberating an electron in the process. For this and other such changes, the "weak interaction" force was proposed. This force was also very short-range, but it was considerably weaker than either the strong or the electromagnetic interactions.

A fourth interaction is the "gravitational interaction," but this is so exceedingly weak that it plays no measurable role in the subatomic world, although it is the dominant force when large masses of matter are considered over astronomical distances.

No fifth interaction has ever been discovered, and at the moment it is not expected that any will be found. In terms of the forces that cause subatomic particles to interact, the Universe seems to be in good simple shape.

The subatomic particles can be divided into two groups. There are the massive "hadrons," which are affected by the strong interaction, and the less massive "leptons," which are not. The proton and neutron are each a hadron; the electron is a lepton.

As the twentieth century wore on, it became clear that the neutron, proton, and electron did not answer all questions. There had to be "anti-particles" that resembled the ordinary particles in every respect except that some key characteristic is in an opposite form. There is an anti-electron (or "positron"), just like an electron but positively charged; an anti-proton, just like a proton but negatively charged; an anti-neutron, just like a neutron but with a magnetic field in the opposite direction.

To explain certain subatomic events, a neutrino (and a companion, an anti-neutrino) had to be postulated, and they were indeed eventually detected. They had neither mass nor electric charge.

By 1960 scientists knew of eight leptons: the electron and anti-electron, the neutrino and anti-neutrino, the muon and anti-muon (a muon is just like an electron but is about two hundred times as massive), and a muon-neutrino and anti-muon-neutrino. (The muon-neutrino differs from the ordinary neutrino since both take part in different subatomic changes, but the exact nature of the difference has not yet been worked out.)

In addition there is the photon, the particle-like unit of light that composes radio waves, x-rays, gamma rays, and electromagnetic radiation generally. A photon is exchanged whenever two particles undergo an electromagnetic interaction, so that it is also known as an "exchange particle." Physicists suppose that each of the four interactions has its own exchange particle.

Eight leptons present a not-so-simple picture, but not impossibly so. Physicists could live with it.

Not so in the case of the hadrons. Beginning in the late 1940s, physicists built particle accelerators that produced subatomic particles with greater and greater energies. The proton and neutron were not alone. These accelerators produced many hadrons which existed only at high energy levels and which quickly decayed. The higher the energies available, the more hadrons were formed, until physicists had found hundreds, with no end in sight.

This was unbearable. If there were that many different hadrons, then they had to be made of something still more fundamental, if our intuitive feeling of the simple Universe were to be correct.

In 1953 Murray Gell-Mann came up with the suggestion that all hadrons were made up of "quarks," whose charges were one-third that of an electron in some cases, and two-thirds in other cases. (The name "quark" is taken from James Joyce's *Finnegan's Wake,* where Joyce comes up with "three quarks" as a nonsense verison of "three quarts." "Quark" is thus an appropriate name because three at a time are required to make up protons and neutrons.)

Gell-Mann began by suggesting only two types of quarks, which he called "up" and "down" (or "u" and "d") for purposes of distinction, though the description can't be taken literally. Two d-quarks and a u-quark total up to a zero charge and make a neutron. Two u-quarks and a d-quark total up to a unit charge and make a proton. There are also anti-u-quarks and anti-d-quarks which, properly put together, make up the anti-neutron and anti-proton.

Many of the other hadrons could be satisfactorily built up out of quarks or antiquarks (or, in the case of hadrons known as "mesons," out of one of each). To explain some of the hadrons, however, more massive quarks, "strange quarks" and "charmed quarks," had to be postulated. (These are whimsical names, without real meaning—just physicists amusing themselves. They might be called s-quarks and c-quarks instead.) Particles containing the c-quark (charmed particles) were first detected only in 1974.

There seem to be analogies between leptons and quarks.

Among the leptons, for instance, there is at the least energetic, bottom level the electron/anti-electron and the neutrino/anti-neutrino. At a higher energy level is the muon/anti-muon and the muon-neutrino/anti-muon-neutrino. There are indications now that at a still higher energy level there is a tau-electron/anti-tau-electron and a tau-neutrino/anti-tau-neutrino. Perhaps there are endless such levels of leptons if we could imagine ourselves going up the energy scale endlessly.

Similarly, among the quarks at the bottom level there are the u-quark/anti-u-quark and the d-quark/anti-d-quark; at a higher energy level, there are the s-quark/anti-s-quark and the c-quark/anti-c-quark.

Physicists are searching for a still more energetic pair, the t-quark/anti-t-quark and the b-quark/anti-b-quark, where the "t" and "b" stand for "top" and "bottom"—or for "truth" and "beauty," depending on how poetic a particular physicist feels.

Again, there may be endless such quark-pairs as we imagine ourselves going up the energy scale endlessly. These different quark-pairs are referred to as different "flavors" of quarks.

As one goes up the energy scale, the leptons increase in mass faster than the quarks do. At some very high energy level, leptons and quarks may become equally massive and may perhaps merge into a single type of particle.

The quark theory, unfortunately, is not firmly established. For one thing, quarks cannot be detected as independent particles. No matter how energetically we smash hadrons, none of the quarks that are supposed to compose them are ever shaken loose.

Does that mean that quarks are just mathematical abstractions that have no concrete reality? (After all, ten dimes make a dollar, but no matter how you tear a dollar bill, no dime will ever fall out of it.)

One theory is that the attractive forces holding individual quarks together within a hadron grow stronger as they are pulled apart. That would mean that any force serving to pull the quarks apart would quickly be overwhelmed as the attraction between quarks is increased in the process.

Quarks differ among themselves as leptons do. Leptons carry an electric charge which may be either positive, negative, or zero. Each different flavor of quark on the other hand has something called "color," which can be either "red," "green," or "blue." (This is just a metaphorical way of speaking, and is not to be taken literally.)

Apparently, when quarks get together three at a time, there must be one red quark, one green quark, and one blue quark, the combination being "white." When quarks get together two at a time, then there must be a color and its anti-color, the combination again giving white.

The behavior of quarks with respect to combinations by charge and color is described in a theory called "quantum chromodynamics," abbreviated "QCD."

Quarks interact by means of the strong interaction, and an exchange particle should be involved, one that is analogous to the photon in the electromagnetic interaction. The exchange particle for the strong interaction is called the "gluon" (because it "glues" the quarks together).

The gluon is more complicated than the photon. Charged particles interact by way of the photon, but the photon has no charge of its own. Colored particles interact by way of the gluon, but the gluon itself has color. There are, in fact, eight different gluons, each with a different color combination. Nor can gluons be shaken out of hadrons any more than quarks can.

This is unfortunate. As the number of varieties of quarks and gluons, with all their flavors and colors, increases, and as quantum chromodynamics gets more and more complicated, the whole structure begins to seem less likely, and to need more experimental support. If there were some way in which quarks and gluons could be detected, physicists might have more confidence in quantum chromodynamics.

Even if quarks and gluons can't be shaken out of hadrons, might it not be possible for them to be formed out of energy? Physicists form new particles out of energy every day. The more energy they have to play with, the more massive the particles they form. If they can get *enough* energy, they could form quarks.

With enough energy, they would form groups of quarks of different flavors and colors. Naturally, quarks, so formed, would instantaneously combine in twos and threes to form hadrons. Such hadrons would stream out in two opposite directions (given enough energy), one stream representing the hadrons, the other the corresponding anti-hadrons.

Where would the necessary energy come from? The most energetic particle accelerators that now exist are particle-storage rings that pump enormous energies into electrons in one place and enormous energies into positrons in another place. When the electrons and positrons are both speeding along almost at the speed of light, they can be made to collide head-on and annihilate each other, converting all their mass into energy. The energy of motion plus the energy of annihilation comes up to about fifteen billion electron-volts, and this should be enough to form quarks.

And there have been such experiments, which did indeed produce streams of hadrons and anti-hadrons. The higher the energy, the tighter and narrower the streams are.

But what about gluons? If gluons come off too, we would expect to see three jets of hadrons coming off at angles of 120 degrees, like the three leaves in a three-leaf clover.

To get this, one ought to have higher energies still. A new particle-storage ring was built in Hamburg, Germany, capable of producing energies of up to thirty billion electron-volts.

Even this amount of energy is just barely above the requirement for producing the gluons, so that one would not expect to have a clear three-leaf clover effect. Using the Hamburg machine, one usually got the two jets, but every once in a while there seemed to be the beginning of a third jet, and this was enough to make some physicists feel that the gluon had been detected.

Even if it was, however, it is still disturbing that there are so many flavors and colors of quarks and gluons, and that there is a second group (though a simpler one) of leptons and photons. Can it be that once again the ultimate has receded and that we must ask ourselves if leptons and quarks alike are built up out of still more fundamental particles?

A physicist, Haim Harari, suggests that this more fundamental particle might be called the "rishon," the Hebrew word meaning "first." He points out that if one imagines a T-rishon with an electric charge one-third that of an electron and a V-rishon with no charge (together with an anti-T-rishon and an anti-V-rishon), then all the lower-energy leptons and quarks can be built up of rishons taken three at a time.

Still considering how difficult it is to get evidence for the existence of quarks and gluons, the thought of going beyond that to get evidence for still more fundamental particles would be enough to make the most hardened physicist quail.

Even if we succeed, will the "rishon" or something like that be the answer? Or is there no answer to the question of what the Universe is made of? Do the ultimate particles we search for recede endlessly and mockingly as our instruments and theories become more subtle, luring us always on to one more step . . . then one more step . . . then one more step . . .?

Afterword: This essay was written in February 1980. Since then, the suggestion concerning the "rishon" has not made any headway, and I have seen no mention of it for a long time.

I say in the essay that "no fifth interaction has ever been discovered, and at the moment it is not expected that any will be found." Within this last year, a fifth interaction, weaker than any of the others, has been postulated, but so far, most physicists remain completely skeptical of it, and (for what it's worth) I am skeptical, too.

27

SCIENCE AND
SCIENCE FICTION

The next eleven essays leave the matters of science and society that have largely occupied me for the first twenty-six essays. Instead, these that are upcoming deal with my various obsessions. First of all, there is science fiction, which I have been industriously writing for nearly half a century, and then a few other things.

I have no real excuse for this except that:

1. As I told you in the introduction, this is a miscellaneous collection.

2. The essays exist and have been published and therefore might as well go into a collection, and . . .

3. My publishers tend to humor me and let me do as I please—and I please to do this.

Two astronauts floated free in space in 1984. They were not tethered. They left the ship and returned to it on their own.

Those of a certain age were reminded of Buck Rogers cartoon strips of the 1930s. It was done there. Those who are still better acquainted with science fiction (s.f. in abbreviation, *not* the barbarous "sci-fi") know that science fiction stories dealt with such things even earlier. Hugo Gernsback, the editor of the first s.f. magazine, wrote about reaction motors that could keep men in flight, in the atmosphere as well as in space, as early as 1911.

The free-float space flight, therefore, has followed science fiction only after nearly three-fourths of a century.

The trip to the Moon itself, carried through successfully for the first time in 1969, follows by over a century the first attempt to describe such a trip with some attention to scientific detail. That was Jules Verne's *From the Earth to the Moon* published in 1866.

Verne used a giant cannon, which was impractical, but the first mention of rocketry in connection with a Moon flight was by Cyrano de Bergerac (yes, the fellow with the nose—who was also an early science fiction writer). His story of a flight to the Moon was published in 1655, thirty years before Isaac Newton demonstrated theoretically that the rocket could carry people to the Moon, and that *only* the rocket would do it. So rocketry lags three centuries behind science fiction.

And where are we to go now? Are further advances in space going to overtake and outstrip science fiction once and for all? No danger!

NASA is now talking of a permanent space station occupied by a dozen or so astronauts in shifts. Gerard O'Neill suggested space settlements, virtual cities in space, holding ten thousand people or more apiece, back in 1974, and in the early 1960s serious suggestions for the building of a solar power station in space were first advanced.

However, back in 1941 Isaac Asimov (that's the fellow writing this piece) published his story "Reason," which describes in some detail a solar power station in space. In 1929 Edmond Hamilton wrote "Cities in the Air" in which, as the title implies, the cities of the Earth declare their independence of Earth's surface, and loft themselves into the atmosphere. That was not quite space, but it paved the way for James Blish, who began his stories about cities in space in 1948.

By the time space stations and power stations are built in space, they will have lagged behind science fiction for at least half a century.

Once we have space stations, of course, we will be able to take longer trips. Making use of mines on the Moon (sometimes mentioned in space travel stories of the 1930s) and the special physical conditions of space, spaceships much larger than those practical on Earth's surface can be built and launched. They will be manned by people accustomed to space travel and space surroundings, and who are psychologically fit (as Earth-bound people are not) to spend months or even a few years in travel.

Yet that is old stuff in science fiction. Space travellers have wandered all over the Solar System since the 1920s. In 1952, in my own story, "The Martian Way," I described, with some attention to scientific detail, a flight to Saturn. In the course of it I described what later came to be called a "space walk," *with* a tether. My description of how it felt turned out to be close to what astronauts reported a dozen years later. So no matter where we go in the Solar System, the astronauts will be following science fiction heroes, some more than a century old.

And after the Solar System comes the stars. Even the nearest stars would take years and decades to reach—even if we went at the speed of light, which theoretically is the fastest possible. If we went at reasonable less-than-light speeds, it would take generations to reach any of the stars.

There are strategies that might be used, but every one of them has been tried out in science fiction already. Arthur Clarke's *2001: A Space Odyssey* placed astronauts into frozen hibernation to wait out the long journey. Paul Anderson's *Tau Zero* was one story of many to have astronauts take advantage of the slowing of time at high speeds. And back in 1941, Robert Heinlein, in his story "Universe," described a large ship, a small world in itself, that was taking a trip of countless generations, ready to spend millennia to reach its starry destination.

If any of these strategies are eventually taken in reality, the astronauts will probably be centuries behind science fiction.

Will faster-than-light travel be possible? I have a tendency to say, "No!"— but it is unwise to be too categorical in such things. In any case, back in 1928

Edward E. Smith wrote the first story of interstellar travel using faster-than-light speeds, *The Skylark of Space*. He invented the inertialess drive, which is probably impossible, and which in any case would only achieve light-speed, nothing more, but the principle remains. If faster-than-light speeds are devised, they will be far behind science fiction.

Has science fiction, then, anticipated everything? Oddly enough, no! Some very simple things were missed completely by the ingenious s.f. writers.

For instance, they concentrated on space flight under direct human control and never realized what could be done by remote-controlled probes. They foresaw computers but missed their true role in spaceflight. (In 1945, in my story "Escape," I had an enormous computer help solve the problem of interstellar flight, but neither I nor any other writer foresaw the microchip and the tininess and versatility that computers could reach, and how essential they would become to the piloting of the space shuttle, for instance.)

We missed all the variety of things that could be done in space. We had trips to the Moon, but we had no weather satellites, no communication satellites (though Arthur Clarke first described them in a *non*fiction piece in 1945), no navigation satellites. . .

Oddly enough, although we foresaw landings on the Moon, and also foresaw television, no science fiction writer ever anticipated reality by describing a landing on the Moon that was watched by hundreds of millions of people on Earth *as it happened*. (The comic strip *Alley Oop* had something of the sort, however).

The conclusion is this. It is unlikely that science and technology, in their great sweeps, will ever outstrip science fiction, but in many small and unexpected ways there were and will undoubtedly continue to be surprises that no science fiction writer (or scientist, either) has thought of.

It is these surprises that are the excitement and glory of the human intellectual adventure.

Afterword: My statement that no science fiction writer had anticipated the fact that the first landing on the Moon would be watched on television as it happens turns out to have been wrong. Sharp-eyed and long-memoried readers wrote to give me several cases where it was foreseen, one written nearly sixty years ago. One of the foreseers was Arthur Clarke, who apparently had forgotten that he had foreseen it and also went around saying that no science fiction writer had anticipated the fact.

28

THE DARK VISION

Herbert George Wells (1866-1946) was a Socialist. He hated Great Britain's class structure, the calm assumption of its upper classes that they were "gentlemen" and "ladies" and, therefore, the "betters" of others by inalienable birthright.

He hated the manner in which this assumption of "knowing one's place" imprisoned people in a long and detailed hierarchy of "upper" and "lower" with everyone truckling to those above them and making up for it by stamping on the faces of those below them.

And it made no difference whether someone showed native ability, drive, ingenuity, industry, and a dozen other admirable qualities. It made no difference whether one became learned or even rich. You were what you were born, and the dirt-poor ignoramus who was a "gentleman," and had the manners to prove it, knew himself to be the better man, and behaved with calm, unselfconscious insolence toward the learned professor or rich manufacturer who was of the "lower classes."

Wells had good reason to hate all this, for his parents had been servants who became small shopkeepers, a miniscule sideways step that might or might not be up or down. Either way it left Wells not much above the status of a laborer. (Being a laborer was the lowest status an Englishman could have. Only Irishmen, Jews, and assorted foreigners were lower.)

It was all the worse since Wells was apprenticed to a draper when he was fourteen, broke away to be an unsuccessful teacher, obtained his degree with the greatest of difficulty, and would have remained completely unknown and un-valued if he had not turned out to have an amazing writing talent and the ability to write science fiction of a quality that had never before been seen (even counting Jules Verne, who suffered under the devastating disadvantage, in any case, of being unEnglish).

Wells grew rich and famous, but remained as much a member of the lower classes as ever in the eyes of his "betters." What was worse, he remained so in his own eyes.

Why, then, shouldn't he hate the class system and labor (as he did all his life) to persuade people to overthrow it, and to set up a system that rewarded brains and industry indiscriminately, and to abolish forever the enormous gulf in material welfare between those born into different classes?

One problem interfered, however. Wells's inability to shake his own sense of inferiority to the aristocracy was matched by his inability to shake his own sense of superiority to the laboring classes he considered beneath him. He was as conscious of being someone's "better" as any Duke could be. More so, perhaps, because the Duke took it for granted and didn't have to demonstrate it defensively at every step, whereas Wells did.

Consequently, Wells was never a true revolutionary. Not for him was there a doctrine of "Workers, arise!" Not for him was there a concept of the downtrodden masses striking off their own chains and turning the world upside down.

No! To Wells, it might be true that the upper classes would never have the decency to allow a cruel system to be changed, but neither could one trust the lower classes to have the decency to change it with efficiency and intelligence.

He may have been right in both respects, but his solution was to have the unearned prestige and power wrenched from the upper classes by the force of reason and persuasion, and then bestowed on the lower classes, with adequate safeguards, by a paternalistic and benevolent intellectual aristocracy.

Golly, that sounds good. The only trouble is that you can search through history and you won't find an example of that working. Revolutions often start with leaders who are intellectual, well-meaning theorists, but generally they are quickly torn apart by the fire-eaters who won't or can't wait.

You can see Wells's views in the novels he wrote. Here we might consider two of his social documents, two dark visions, in the form of novels (and powerful, intelligent, and interesting ones, too). These are *The Island of Dr. Moreau*, first published in 1896, and *A Story of Days to Come*, published in 1900.

A Story of Days to Come is intensely English; the rest of the world scarcely counts and is rarely even mentioned. The tale is set in the year 2100, more or less, and there are two important factors. First, the world has advanced technologically in the two centuries since Wells's own times; and second, the world has *not* advanced socially. Rather, it has retrogressed.

The advance of science has only made it easier for the upper classes to oppress the lower classes, and has also increased the fearsomeness with which the lower classes might take matters into their own hands if they get the chance.

Wells is persuasive, all the more so since nearly half the two centuries he looked across has now passed, and we can judge the level of correctness he displayed.

We find that, socially, Wells is a rather accurate prophet. Despite all the technological advances we have experienced since his day, the gap between rich nations and poor nations is worse than ever, and the gap between the rich and poor within a nation continues to be at least as bad. The "haves" are as reluctant to share, and the "have nots" are as avid to grasp, when they have the chance, as ever; and both are as immune as ever to the music of sweet reason.

Remember that the story was written some decade and a half before World War I and the Russian Revolution, and nearly four decades before World War II and the Nazis, so that neither communism nor fascism could be used as models. Nevertheless, the tale manages without them, and it ends up being much better, in my opinion, than George Orwell's vastly overrated *1984*.

A Story of Days to Come is a soap-opera of the future, and a moving one. It is the story of a pair of lovers, ground, like Wells, between the millstones of being not good enough for the upper classes and being too good for the lower classes. That it is done with intimacy, and a touch of humor as well, makes it more believable and poignant.

Particularly interesting, of course, are the technological innovations that Wells foresees. Wells has his limitations in this respect, and it would be the easiest thing in the world to go through the tale and sneer at his mistakes and insufficiencies with all the clear vision of hindsight. One must not. Wells is a genius of foresight compared to anyone else. If *he* could not see with perfect clarity, it is not because he was insufficient, but because *no one* could.

The lesson to be learned is that all human beings, in trying to assess future technology, are bound and shackled by their present. Wells's superb imagination could no more free him from being bound to what existed in 1900 than his pounding Socialism could free him from his own class consciousness.

Wells was writing his book just as Freud was writing *his* first great book, but Wells could not see Freud's work coming. Wells imagined various psychosomatic illnesses and anti-social behavior being treated by hypnotism in the world of 2100—but he didn't write one word about psychiatry.

Also, Wells wrote this book just as Planck revolutionized physics with his quantum theory, and just a few years before Einstein revolutionized it in another way with his theory of relativity. Wells, of course, foresaw neither (how could he?), so the world of 2100 that Wells gives us is neither quantized nor relativistic.

Again, Wells foresees the end of books, but does not see them being done in by television and the computer. No, indeed, he finds the telephone sufficient for the task. The telephone was a quarter century old when the book was written, but Wells could not manage to get beyond it.

Wells foresees aircraft, but not in the form of what we would today call airplanes. He does speak of "airplanes," but what he describes are zeppelins, the first of which flew in 1900. The true airplane did not fly till 1903, so that invention of science and technology remained beyond him.

Conclusion: it is hard to underestimate the speed with which humanity improves its social habits, since they sometimes appear not to improve at all; and it is hard to overestimate the speed with which humanity improves its technology, since even the genius of H. G. Wells fell short in imagining how far and fast it would come.

The Island of Dr. Moreau is much darker than *The Story of Days to Come*. It is a bitter tale of beast made into quasi-men, but unable to maintain the imposture. Taken literally, it is the story of a distasteful scientific experiment gone wrong.

View it, however, in the light of Wells's social views, and it becomes an allegory of the human adventure. Are not *we* beasts who have made ourselves "human"? Or, in the sad cry of the book: "Are we not Men?"

Those who say it in the book are *not,* alas. Can it be that we ourselves, two-legged and five-fingered by birthright, are not men either? Can it be that though we call ourselves human, the ancestral unbrained peeps through? In the last chapter, when Prendick returns, his view of humanity, as sickly negative as Gulliver's after his fourth and climactic voyage, gives Wells's clear and pessimistic answer.

And remember, the story was written nearly twenty years *before* the beginning of the cycle of world conflict, which, once it did begin, showed human beings behaving far worse than Dr. Moreau's innocent beast-victims.

As for *The First Men in the Moon,* first published in 1901, it is a classic of science fiction, absolutely the best and most sensible interplanetary tale told up to that time.

It has its digs at social greed. Bedford (the first person narrator) manages to reveal himself as a rather unlovely person. He is interested in Cavor's scientific discoveries only for the money and power he thinks he can get out of them. He is more fascinated by the gold he finds on the Moon than by anything else. He is a whiner and complainer, easily in despair and easily in rage—in short, he embodies everything that prevents the social machine from functioning well.

(Cavor, the scientist, on the other hand, is intelligent, brave, and idealistic, but so hopelessly impractical that he is surely not to be trusted with the social machinery either.)

There are also scientific shortcomings. Wells has the unusual idea (for the time) of postulating a substance opaque to gravity. Insulate yourself from gravity, and the Earth moves out from under you. By shielding and unshielding yourself in the proper way with this substance, one can even maneuver oneself all the way to the Moon.

It's a beautiful propulsive mechanism—extremely cheap once the gravity-opaque material is developed, since no energy is required; and there is no acceleration either, since you don't move—the Earth and Moon do.

The catch is that a gravity-opaque material isn't in the cards. It isn't possible even theoretically, if Einstein's general theory of relativity is correct, and if gravitation is looked upon as a geometric distortion of space-time. (Jules Verne, who in his last years was furiously jealous of Wells's success, sneered at Cavorite and defied Wells to produce a sample. Science fictionally, that was a foolish attitude to take, but Verne's instinct was correct. There was no such thing as Cavorite and there could not be—as far as we know today.)

Still, *The First Men in the Moon* was written fifteen years before general relativity, so Wells's sin is a forgiveable one.

It is amusing, by the way, that Wells drags in very recent scientific advances in his discussion of Cavorite—a good science fictional technique. He mentions "those Roentgen rays there was so much talk about a year or so ago" (as though x-rays were a passing fad), and Marconi's "electric waves" (we would say "radio waves.") Most interesting of all, is Wells's careful hint that one of the substances added to the "complicated alloy of metals" that constituted Cavorite was "something new." It was nothing else than "helium," first discovered in 1898. Wells carefuly did not mention (perhaps he had not as yet learned) that helium was totally inert and, of all substances, the least likely to incorporate itself into the Cavorite alloy.

Wells's picture of a Moon with air, water, and life is at variance with the truth of the matter.

In this, I don't mean that we *now* know better. Of course, we do, but that doesn't count. The point is that the astronomers in 1901, when the book was published, knew better.

Wells's supposition is that the Moon has an oxygen-containing atmosphere which freezes during the Moon's long night. Well, the lunar night is colder than Antarctica, but it isn't cold enough to freeze oxygen and nitrogen.

That can be forgiven, for in 1901 there might have been an argument as to just how cold the Moon got during its nighttime period. *But* if the airy snow remained for any appreciable time, astronomers would be able to see bright spots on the Moon shine then fade as the Sun passed. Needless to say, scientists have never seen such spots.

Furthermore, as the ice melted to produce enough vapor to turn the sky a faint blue, and as vegetation grew rapidly, astronomers on Earth would surely detect changes. They would find the outline of the craters blurring somewhat; they might make out faint mists or clouds; they might detect some color changes as vegetation turned first green, then brown. And none of this has ever been detected.

But what the heck. Why spoil a good story?

Even though we know better, Wells describes the Moon with such verve, such brilliance, such inner consistency, and such excitement, that there is a most willing suspension of disbelief. Even if it isn't true, it *should* be true.

Let's summarize: These stories are over eighty years old, but however they have been outmoded, or reinforced, one thing hasn't changed at all. They are still terrific stories.

29

THE LURE OF HORROR

In 1974 a book of mine was published which was a detailed annotation of John Milton's great epic, *Paradise Lost*.

In preparing it, I noted what seemed to me to be an interesting thing, something that was not emphasized in the commentaries I had seen.

Here were bands of angels in countless hordes, all living in unimaginable bliss in the company of a perfect and all-beneficent God, and yet one-third of them rebelled and fought desperately against the remainder. The rebels were defeated, of course (of course!), and were hurled into the unimaginable tortures of hell, and yet not one of them (not one!) rebelled against Satan for having led them into this ultimate catastrophe. Not one expressed remorse or contrition. Not one suggested trying to truckle to God and seek pardon.

Why is that?

It is no use to fall back upon "sin" and say that sin had seized control of the fallen angels. Sin is just a one-syllable word that describes the situation. *Why* did sin seize control? How could it find entry under conditions of absolute perfection?

Nor is it enough to say that this enigma was just the product of Milton's eccentricity. The epic poem could scarcely have been so universally praised if intelligent readers did not find something deep and satisfying in Milton's analysis of the situation.

So I looked, and here is what I found. In Book II of *Paradise Lost* the fallen angels gather in council to decide what must be done to remedy their situation. Satan throws the floor open to discussion and several of the leading rebels state their disparate views, after which Satan comes to a rational decision, taking on the burden of its implementation himself. In Book III, there is a council in heaven—but only God speaks, and the role of the angels consists of the equivalent of an endlessly repeated "Yes, Lord."

Again, after the council in hell is completed, the rebel angels proceed to amuse themselves as well as they can within the confines of hell. Some engage in war games, some in athletics, some in music, some in art, some in philosophy, some in exploration of their new domain. In short, they behaved as though they were rational beings. In heaven, the unfallen angels did not seem to have any role but to sing the praises of God and to follow his orders.

I could not help but think that an existence so limited as that described in

Milton's heaven, however much it was *called* bliss, would produce that serious mental disorder called "boredom." Surely, given enough time, the pains of boredom, of dissatisfaction with an absolutely dull and eventless life, would gather in intensity to the point where any being with a rational mind would see it as worse than any conceivable risk that accompanied the attempt to end it.

So Satan and his angels rebelled, and when they found that they were thrown in hell as a result, they were dreadfully distressed, but not one of them felt like returning to heaven. After all, there was self-rule in Milton's hell, and as Satan said, in the single best-known line of the epic: "Better to reign in Hell, than serve in Heav'n."

In short, then, one of the things that *Paradise Lost* explains, in very satisfying manner, is the lure of "horror" upon all of us. That this lure is very real is beyond doubt.

Consider the tale of adventure and daring deeds. That is a genre undoubtedly as old as *Homo sapiens*. Around Stone Age campfires, men surely told tales of danger and heroism that the listeners found not only entertaining but filled with social meaning. The heroes of the stories, whether those stories were exaggerated truth or outright fiction, were role-models that made men more daring in hunt and more enduring in adversity. Such things therefore contributed to the advance of the tribe and of humanity in general.

Yet there is no question but that in these tales, those elements which encourage fear and dread are more strongly emphasized than would seem necessary, and that, indeed, the popularity of such stories is increased in proportion to the efficacy with which such fear and dread is introduced.

The oldest surviving and the most continuously popular adventure story in the Western tradition is Homer's *Odyssey*. Does anyone doubt that the most popular portion of the epic are Books Eight to Twelve, Odysseus's tale of his wanderings through the terror-lands of the (then-unknown) western Mediterranean? (Indeed, I suspect that some who know only what they have heard of the *Odyssey* imagine that this section comprises the whole of the tale.) And, of Odysseus's tale, surely the most popular item is also the most terrifying, his encounter with the one-eyed cannibalistic giant, Polyphemus.

The tendency has continued through all the twenty-eight centuries or so since the *Odyssey* was composed. We find ourselves fascinated with stories that inspire suspense, that induce in us increasing uneasiness. We read, without danger of satiation, adventure stories, mystery stories, gothics, and, always popular, those tales which distil out of all this just those elements which most efficiently arouse fear and dread. These are the so-called "horror stories," which confront us with the nearly unendurable.

Why should we read these when none of us wants to be thrown into real-life situations involving fear and dread? We don't want to face wild beasts, or spend a night alone in a haunted house, or have to run from a homicidal maniac. Of course, we don't want to. But a life filled with peace and quiet becomes dreadful and would drive us to these things if we did not find some way of exorcising boredom (as the rebellious angels could not).

We read about events which take place in the never-never land of fiction, and for a while we are absorbed in them and feel the emotions they give rise to. We experience them vicariously without *really* being in danger and can then return to our peace and quiet with the risk of having to "break out" allayed.

This need to escape from the disagreeableness of the potentially unbearable quiet life would, in fact, seem to be deeply ingrained in human nature. Where is the child that does not love a ghost story? Where is the popular children's folktale that does not involve witches and spectacular dangers of all kinds? It may induce nightmares and persistent fear, but even that seems to be preferable to a life of utter peace. Children seem to agree with the rebellious angels; all of us do, in fact.

The modern horror story begins with *The Castle of Otranto,* by Horace Walpole, published in 1764. The scene is an unlikely medieval castle with an impossible architecture that leaves room for interminable secret passages, and a plot that includes ghosts and supernatural influences of all kinds.

Many followed Walpole's lead, and the genre reached its peak in the nineteenth century with Edgar Allan Poe, his predecessors, and his many imitators. Some of these stories, such as Poe's "The Tell-Tale Heart," have never lost their popularity, or their efficacy. Most, however, have, for fashions in horror have changed (as all fashions do).

Might we say that change has come because we no longer need ghosts and the supernatural in the twentieth century, where the horrors of war and the chances of universal destruction are so much greater than they ever were in preceding centuries?

Indeed? Surely, the unknown still terrifies. With advances in technology, we call upon vision to make real what, earlier, only imagination could embody. The twentieth century horror movies such as *The Exorcist* and *Poltergeist* are new versions of the nineteenth-century horror tale, and the lure remains.

30

MOVIE SCIENCE

Think of the hundreds of millions of people the world over who have seen "Star Wars," "E. T.," "Close Encounters of the Third Kind," "Star Trek: the Movie," "Superman," and all the other blockbuster science fiction films. Each of them portrays worlds in which some aspects of technology are advanced far beyond anything we have today. (That's why we need "special effects.")

The question is, though: How many of these aspects of future technology have a reasonable chance of actually coming to pass in the foreseeable future? Are the movies, to some extent, being prescient, or are they indulging in fantasy?

In considering the answer to that question it would be unfair to insist on accurate detail in order to give a movie satisfactory marks, for that would be asking the impossible.

In 1900, for instance, some bold spirits were thinking of the possibility of heavier-than-air flying machines. If we imagine a motion picture having been made in 1900, with the action set in 1984 and large flying machines capable of supersonic flight playing an important role, that motion picture would surely be considered to have represented successful futurism. However, what are the chances that the 1900 special-effects people would have put together a structure that would actually have resembled the present-day Concorde? Vanishingly small, surely.

When we see films, then, in which spaceships routinely exceed the speed of light, it isn't fair to ask whether the superluminal (faster-than-light) ships of the future will really look like that. Instead, we should ask whether superluminal vessels, of any kind whatsoever, are possible.

In our Universe, under any conditions even vaguely like those with which we are familiar, superluminal speeds would *not* seem to be possible—so that Galactic Empire epics must be considered fantasies. If we are bound by the speed-of-light limit, it will take years to reach the nearest stars, thousands of years to span our own part of the Galaxy, a hundred thousand years to go from end to end of our Galaxy, millions of years to reach other galaxies, billions of years to reach the quasars. Relativity, which tells us this, also tells us that astronauts moving very near the speed of light will experience very little sensation of time passage, but if they go to the other end of the Galaxy and back under the impression that only a short time has passed, two hundred thousand years will nevertheless have passed on Earth.

Some people rebel at the speed-of-light limit. Surely there must be some way of evading it.—Sorry, not within the Universe.

Why not?—Because that's the way the Universe is constructed.

After all, if you travel on the surface of the Earth, you cannot ever get more than 12,500 miles from home. Nothing you do, on the surface, can get round that. If, at a point 12,500 miles from home, you move in *any* direction, you move closer to home. That would seem paradoxical and ridiculous if the Earth were flat, but it is the consequence of the Earth being a sphere of a certain size, and there's nothing you can do about it—except get off the surface of the Earth. If you go to the Moon, then you are 237,000 miles from home.

Just as the Earth is surrounded by a vast Universe into which we may escape from our bondage to the surface, so it might conceivably be that our Universe is surrounded by a vaster "hyperspace" of broader and more complex properties than ordinary space, and within which the speed-of-light limit might not hold. Science fiction uses hyperspace routinely for interstellar spaceflight but is never explicit, naturally, on the means of moving into it and back out of it.

Some physicists have speculated on the existence of "tachyons," particles that have the property of always going faster than light and moving ever faster as their energy decreases. They, and the space in which they exist, may represent an analog to science fiction's hyperspace. However, such tachyons have never yet been detected and some physicists argue that they are theoretically impossible because they would violate causality—the principle that a cause must precede in time the effect it induces.

Some physicists have speculated that within a black hole conditions are so radically different from ordinary space that the speed-of-light limit might not hold. A black hole might therefore be a kind of tunnel to a far-distant place in the Universe. There is, however, nothing beyond some vague theoretical considerations to support this and most physicists do not believe it to be so. Besides, there is nothing we can yet conceive of that would allow human beings to approach a black hole, take advantage of these phenomena, and still stay alive.

On the whole, then, faster-than-light travel must be left to science fiction for the foreseeable future, and possibly forever.

The same can be said, even more strongly, for another staple of science fiction plots—time-travel. The first writer to make systematic use of a time-travel device (as opposed to being taken into the past, let us say, by the Ghost of Christmas Past, or by a knock on the head of a Connecticut Yankee) was H. G. Wells in his 1895 tale "The Time Machine." Wells's rationale was that time was a dimension like the three spatial dimensions of length, width, and thickness, and could therefore be travelled through similarly, given an appropriate device.

Einstein's theory of relativity does indeed treat time as a fourth dimension, but, alas, it is *not* like the other three in nature, and is treated differently in relativistic mathematics. There is reason to believe that not only is time-travel impossible at the present level of technology, but that it will be forever impossible.

We can see this without delving into relativity. Clearly, time-travel will destroy the principle of causality, but beside this, one can't deal with time-travel in any way without setting up paradoxes. One can't move into the past without changing all events that follow the point at which one arrives. Time-travel therefore implies the existence of an infinite number of possible "time-lines," and it becomes difficult, or perhaps impossible, to define "reality." (I made use of this view in my novel *The End of Eternity*.)

An even simpler point is that time-travel is inextricably bound up with space-travel. If one moves back or forward one day in time and is still on Earth, one has to take into account the fact that Earth has moved a great distance during that day in its journey around the Sun, and an even greater distance in its accompaniment of the Sun in *its* journey about the center of the Galaxy, and perhaps a still greater distance along with the Galaxy in *its* motion relative to the Universe generally.

Remember, too, that this sets a definite limit to how fast one can move through time. It would take a whole day to move five years into the past, or future, since to do it in less time would certainly involve faster-than-light travel through space. And where would the energy come from?

Less commonly used, but still very convenient, is the notion of anti-gravity, and its inverse, artificial gravity. If we could have anti-gravity, spaceships could move off the surface of Earth without having to use any more energy than is required to overcome air resistance. With artificial gravity, human beings could live comfortably on any asteroid and make sure it would hold an atmosphere indefinitely.

Unfortunately, however, if Einstein's theory of general relativity is correct, gravitational fields are the result of the geometrical distortion of space induced by mass, and by nothing else. It cannot be blocked or simulated except by the use of masses as great as those which produce it normally. In other words, you can't neutralize the effect of Earth's gravity on a ship on its surface except by holding another Earth immediately over it; and you can't give an asteroid the gravitational pull of the Earth except by adding an Earth-sized mass to it. There does not seem to be any way out of this dilemma.

What about telepathy or, more generally, mind-control? This is not inconceivable. The neurons of the brain work by means of tiny electrical currents which produce a tiny electromagnetic field of incredible complexity. There might be ways in which such a field could be detected and analyzed, or in which the field of one person could be imposed upon that of another in such a way that the second would carry out the will of the first.

It wouldn't be easy, for in the brain there are 10,000,000,000 neurons all firing their tiny currents. The problem would be similar to hearing ten billion people (twice the population of the Earth) talking simultaneously in whispers and trying to extract definite remarks from the medley. Perhaps, though, it might be done; perhaps certain individuals have the natural ability to do it; and, even if not, perhaps the ability can be gained by way of some artificial device.

It doesn't seem likely, but one hesitates to say it is flatly impossible.

Of course, it is not only the big things that run counter to the possible, or even the plausible. Sometimes a special effect that seems perfectly reasonable is simply out of the question in reality.

In "The Empire Strikes Back," for instance, a ship maneuvers its way through a swarm of tiny asteroids at breakneck speed. In "The Return of the Jedi," ground vehicles maneuver their way through the closely spaced trees of a dense forest, again at breakneck speed. The result is a breathtaking "carnival ride" in each case, since the human reaction time is actually so slow that neither the asteroids nor the trees can possibly be avoided at the given vehicle speed for more than a fraction of a second. The audience, facing instant destruction but surviving, is understandably thrilled and delighted.

Of course, it is not human reaction time that is being depended on for safety, but the far faster reaction time of advanced computers. That is clear even though I don't recall either movie making reference to the fact.

However, that is not enough. A vehicle under the control of an adequate computer would have to make very rapid veers and swerves in order to escape destruction; and that the audience sees it doing so is indeed essential to the thrill. There is, though, something called "inertia," which is the basis of Newton's first law of motion and which nothing in Einstein abolishes.

Though the vehicle bucks and swerves, those contents which are not integral parts of the framework (including the human passengers) tend, through inertia, to remain moving in a straight line. The human beings are therefore slammed by the vehicle at every swerve, and it wouldn't be long before the humans were smashed to a pulp. Avoiding the asteroids or trees by that kind of computer-directed energetic swerving achieves the same effect, and very quickly, as *not* avoiding them.

And yet science fiction is not merely a litany of the impossible. The almost ubiquitous robots and computers of science fiction in the past are now beginning to turn up in reality. Even robots that closely resemble human beings, though they may not show up for some time yet, are not on the same level of unlikelihood as faster-than-light travel.

When I started writing science fiction in the 1930s, a great many things that are now commonplace, including nuclear power, trips to the Moon, home computers, and so on, were given exceedingly little chance of coming to pass within half a century by anyone except the most optimistic science fiction writers. There is much that remains possible, including an infinite number of items that perhaps very few of us even conceive of at present.

So—who knows what wonders lie ahead?

31

BOOK INTO MOVIE

When a story, with its plot and its characters, appears in more than one form, you tend to be imprinted by the first form in which you experience it.

I read Hugh Lofting's "Dr. Dolittle" novels beginning at the age of ten, and I read them over and over. When the movie *Dr. Dolittle* was made I absolutely and resolutely refused to see it. Why? Because Dr. Dolittle in the book, according to description and illustration, was a short and pudgy fellow of lower class appearance. And who played him in the movie? Rex Harrison, the well-known, tall, thin aristocrat. I refused to pay money for such a travesty.

On the other hand, I encountered Mary Poppins first in the movies and fell in love with Julie Andrews, Dick Van Dyke, and all the rest. When my dear wife, Janet, tried to get me to read the Mary Poppins *books* of which she had every well-thumbed title, I drew back in horror. The Mary Poppins illustrated there didn't look at all like Julie Andrews. Janet, however, is very forceful, and though I reared, bucked, and kicked, I finally read the books and fell in love with them, too. But I insist that there are two stories, the Mary Poppins/movie and the Mary Poppins/book, and that they have nothing to do with each other.

And that is, in my opinion, a key point to remember. Books and movies are two different art-forms. The stage is a third. And even upon the stage the same play, presented as a musical comedy and as a non-musical comedy, might achieve contradictory effects.

Pygmalion, as Shaw wrote it, and as Lerner and Lowe rewrote it, is in some ways the same play. The musical retains all of the plot and much of the dialog that Shaw originally wrote and is a faithful adaptation indeed. And yet—when we watch the musical, we must be prepared for, and forgive, the artificiality of having the characters break into song on the slightest pretext, with a complete orchestra appearing from nowhere. Shaw didn't have to struggle with *that.*

And we accept the musical convention. In fact, whenever I see the original play and here Eliza say "Ow-w-w" and Higgins respond with "Heavens, what a noise!" I always have the feeling of stepping off a curb I didn't know was there, because what I expect at that moment is to have the orchestra strike up while Higgins launches into the non-Shavian: "This is what the English population calls an elementary education."

Consequently, when a written story is converted into a movie, it is useless to complain that the movie isn't true to the book. Of course it isn't. It couldn't be.

You might as well complain that a stained glass representation of "Mona Lisa" didn't catch all the nuances of the painting, or, for that matter, that da Vinci didn't manage to catch the fire and gleam of stained glass.

In fact, if some moviemaker, anxious to lose his investment, were to make a movie that paralleled a printed story *precisely,* you probably wouldn't like it, for what is good on the printed page is not necessarily good in the screened image (and vice versa).

This doesn't mean you can't make a bad movie from a good book. Of course you can, but that would be because the movie fails in its own terms, and not because it is untrue to the book. It is also possible for a movie to improve upon a book. *Oliver Twist* is not one of my favorites, but I am thoroughly delighted with the movie *Oliver.*

Why are books and movies so different? Books are a series of words. They are altogether language. Half a century ago and more, books were commonly illustrated—and these images sometimes vied with the printed word in importance (a notable example being *Alice in Wonderland*). Book illustration has now mostly disappeared, so only the string of words remains. Therefore, for a book to be successful that string of words must be well done, must catch the readers' minds and emotions. A bare string of words must substitute for image, sound, intonation, and everything else.

The movie, on the other hand, works very largely with image. Words exist in the form of dialog; there are sound effects and musical accompaniment, but image is primary.

In some ways, the movie image is a much more subtle tool than the words in the book. An effect that can only be created by a paragraph of description can be caught in film by a moment when a fleeting expression is shown, such as a gesture of a hand or a sudden appearance of a knife or a clock—or almost anything.

And yet in other ways words are so pliable, so easy to bend into a flash of irony and wit, so successful in producing long satirical tirades, so subtle in revealing character.

Naturally, in order for each art-form to do its thing well it must emphasize its strengths and slur over its weaknesses, and the result is two different stories— ideally, each wonderful.

There is a second difference between books and movies that is not artistic in origin but economic. A movie is much more expensive to produce than a book is.

For that reason, a book can make money if it sells as little as five thousand copies in hard cover, a hundred thousand in soft cover, and appears in a few foreign editions. Under such circumstances, it will not make the author rich, but at least it will do well enough to make the publisher smile and nod graciously if the writer suggests writing another book.

This means that the writer can aim to please fewer than one out of every thousand Americans. He can aim at a relatively small and specialized group and still make a living and gain success. (For many years I supported myself adequately by pleasing that small group of Americans who were science fiction readers.)

It further means that a writer can afford to be haughty. He can write a book that entirely pleases himself and that does not cater to the general public. He can write a difficult book, a puzzling book—whatever. After all, he need only be read by a fragment of the public to be successful. I don't say that writers scorn best sellers as a matter of course; I merely say that they may do so if they wish.

Not so the moviemaker, who, if he is to get back his much larger investment, must seek an audience in the millions, perhaps even in the tens of millions. Failure to do so can mean the loss of a fortune. To please so many, the movie must be much more careful to hit some common denominator, and the temptation is always present to cheapen the story in consequence.

One common denominator is romance. A story can be written without romance and still do well. A movie without romance finds it much more difficult to be profitable. This means a young woman of improbable beauty is almost sure to find herself thrust into stories where she does not fit very well.

I remember, back in the middle 1940s, seeing a movie version of "The Most Dangerous Game," in which the hero, who must use all his skill, endurance, and intelligence to survive his flight through the wilderness, finds himself saddled with a starlet in evening dress and high heels. I would have roared with laughter had I not been rendered speechless with astonishment.

Finally, a movie is often made from a short story and, if a full-length feature is to be produced, much must be added, so that the screenwriter has all the more impetus to bring creativity of his own to the story. A good screenwriter can use this as an opportunity to improve the story greatly, but, as in all other categories of endeavor, the number of good screenwriters is far fewer than the number of screenwriters.

32

MY HOLLYWOOD NON-CAREER

Here we are, living through a vast boom in science fiction on the big and little screen; one that is beginning to boom in home video, too. And here am I, one of the "big three" of science fiction—and I'm not participating in the boom. I'm standing in a blizzard of hundred-dollar bills and not one of them is hitting me.

How on Earth do I manage that?

Let me tell you, it isn't easy. It takes skill, grit, and determination.

Naturally, I'm going to describe just exactly how I do it. I am not selfish and I know there are many of you out there who would also like to be missed by fame and fortune. Listen to me, my friends, and you too can non-achieve.

There are two cardinal rules you must never forget.

First, refuse to leave Manhattan. (That is, of course, if Manhattan is where you live. If, unlike me, you happen to live in Keokuk, refuse to leave Keokuk.)

Practice the necessary statement with an imaginary telephone at your ear. Pretend a jovial voice is offering you untold sums to involve yourself with some sort of television and/or movie project, and at the crucial moment, when the honey-gurgle in your ear hits a high dollar-note and pauses, you say, "I'm sorry, sir, but I don't travel. All our conferences will have to be on the phone. Failing that, sir, you will have to come to Manhattan (or Keokuk) any time you want to talk to me."

But why stick to Manhattan? Is it just to gain the accolade of failure?

No! Failure may be great but there are practical reasons, too. Travel is a pain in the buttocks and even if it weren't, Hollywood (or southern California in general) is not where you should want to go. It may be that directors, producers, actors, agents, and nymphets are happy there (though I doubt it), but writers aren't. All the Hollywood writers I know are prematurely aged, afflicted with nervous impotence, and they are forever taking quick looks over their left shoulder in the sick certainty that someone is sneaking up on them with a knife aimed at the back.

In Manhattan, on the other hand (and in Keokuk, too, I'm sure), writers are young, virile (or generously proportioned, if female), hearty, and think knives are for slicing bread.

But what if the voice on the telephone chuckles greasily and assures you that it has long been his secret dream to make frequent trips to New York (or Keokuk) and that long conferences with you would make the expense worthwhile.

In that case, there is the second rule. You say, firmly (trying to make your voice sound like a door shutting with a slam and a bolt snicking into place), "I don't do screenplays or television plays."

Does this shock you? Why, for heaven's sake, not?

It's because screenplays and television plays are reviewed by various non-literary personnel—the producers and directors, *of course,* to say nothing of their mothers-in-law and kid sisters. Then there are the actors, the office boys, and the passers-by who have wandered into the producer's office looking for the men's room.

Naturally, every one of them has heavy objections to the screenplay as written by a mere writer. These are relayed to the writer, but only one at a time. It is only after he has rewritten the screenplay to meet the first objection that the second is handed him, and so on.

I know some actual cases of writing in Hollywood. In one case, a particular writer doing a screenplay of a book I know well has been hired on three separate occcasions to work on the same book, just so the producer can have the pleasure of firing him again for the crime of trying to keep the spirit of the book. In the case of another book, a sixth screenplay is in the process of being produced by a sixth writer, and the triumphant report is that the sixth is no closer than the first.

Who needs it all?

Ah, but if I won't leave Manhattan and won't write screenplays just out of a selfish desire for trivial things like sanity and happiness, how do I make a living?

Simple! I write books. (Books!—*Books!*—They're things made up of sheets of white paper, with printing on them, and covers around them.)

I just fill up the white papers and bring it to an editor who sees to it that it's published. Sometimes he asks, in a timid voice, for some minor changes—a comma inserted here, a misspelling corrected there—I usually oblige the gentleman.

I admit that a book doesn't make much money by Hollywood standards, but there's always quantity. I've published 355 books so far and the nickels and dimes mount up. Then, too, there's always the sanity and happiness to consider. Such things may not be as important as money, but, what the devil, I'm easy to please.

Afterword: I foolishly offered the preceding essay to a magazine that was dependent on Hollywood productions. Naturally, it didn't see eye to eye with me when it came to making fun of Hollywood. So I gave it to an amateur science fiction fan magazine. Its first professional appearance is right here.

33

I LOVE NEW YORK

I love New York.

I was brought up there from the age of three (having been born elsewhere, but that wasn't my parents' fault; they didn't consult me). To be precise, I was brought up in Brooklyn, where I obtained that crowning touch of the English language—the Brooklyn accent.

The exigencies of life forced me to spend three years of the 1940s in Philadelphia and the entire stretch of the 1950s and 1960s in Boston. I did well in both places, but I never spent one day in either city in which I was not aware of myself as a New Yorker in exile.

On July 3, 1970, I finally returned to New York, and I will never leave it voluntarily again. Never! I do not even like to leave it temporarily for a short business trip or "vacation."

To be precise once again, I now live in Manhattan, and very close to its geographic center, too, and to me it is like living in Paradise.

There are numerous reasons for this and the first can be expressed as: Walking.

To live anywhere outside Manhattan is to be immersed in sterility. There is bound to be nowhere to go in your immediate vicinity. If there are a few places you lust for, it is automobile time and driving, and your destination, whatever it is, is itself surrounded by sterility. To go somewhere else—back into the car.

Manhattan, on the other hand, is concentrated interest. It is a chocolate chip cookie that is all chocolate chips. If you search for sterility with a magnifying glass you won't find it.

I have but to walk one block west from my abode and I am at the northeastern edge of Lincoln Center. I have but to cross the street immediately to the east of my abode, and I am in Central Park. Half a mile due north is the American Museum of Natural History. One mile northeast across the park is the Metropolitan Museum of Art. Each is the best of its kind in the nation.

Draw a circle a mile in radius around my apartment and within it are at least fifty first-class restaurants of perhaps twenty-five different ethnic persuasions.

Two blocks from my house are two movie houses, one of which shows quality foreign films, and one of which specializes in nostalgia, screening the great classics of the 1930s and 1940s. Four blocks south is a lavish movie theater which shows the latest new releases.

One and a half miles to the south of me is the theater district, the famous "Broadway," declined now from its former magnificence, but still the home of American drama and musical comedy.

And of great importance to me is the fact that anywhere from one mile to two miles south and southwest of my apartment is "midtown," where almost all my publishers have their offices. I can, if I am willing to stretch my legs, walk to Doubleday and Co., to Walker and Co., to Basic Books and Harpers, to Simon and Schuster, to Crown Publishing, all of which have published books of mine. Even Houghton Mifflin, my Boston publishers, has an office in midtown Manhattan.

And if the weather is cold, or drizzly, or you are in a hurry, or are just feeling lazy, you don't have to walk. There is nowhere in Manhattan where you cannot step to the nearest curb and hail a passing taxi almost at once. If you are forced to wait as much as five minutes for one (barring rush hour, or during sudden downpours or other unusual situations), you feel ill-used.

When I can, though, I prefer to walk.

Isn't it dangerous, though, to be walking in New York? Well, the streets are crowded, traffic can be pandemonious, cornices have been known to fall from buildings. There is no place on Earth where it isn't necessary to be careful and have your wits about you, but New York, allowing for the hectic pace of its crowded life, is no more dangerous than anywhere else; probably less so.

Muggings? Crimes of violence?

A whole mythology has grown up about New York, to the point where tourists enter the city feeling they must have bulletproof vests and looking eagerly about for bullet-riddled corpses in the doorways.

Nonsense! There are crimes of violence against people and property everywhere in the United States, and New York is tops in absolute amount only because there are more people here than anywhere else. Study the statistics on a per capita basis and New York is *not* the crime capital of America. Atlanta and Detroit are far more violent and murderous than New York. So are a dozen other metropolitan areas.

Even Central Park, which is supposed to be inhabited only by muggers and other human beasts of prey, is, in actual fact, a bucolic area of peace at most times. I can look at it from the windows of my apartment, and there is no time during the day when it is not filled with joggers and children, and (on warm, sunny days) picnickers and sunbathers.

At night, I admit, I would not willingly enter the park, but that is only because the mythology keeps virtually all citizens out of the park, leaving it, for the most part, to the lawbreakers. In this way the mythology, by its own force, becomes truth.

In fact, not only are the people of New York not especially dangerous to each other, they are one of the chief fascinations of the city. Nowhere in the United States, possibly in the world, is there such a fascinating microcosm of humanity. All nationalities, all languages, all religions, all shades of color mingle and flow through the streets.

One street to my west is Columbus Avenue, another street west is Broadway. Head north on either street and you are entering the fabulous "upper West Side," where the prismatic stream of men and women spreads the geography of the world before your eyes. Columbus Avenue, with the burgeoning number of restaurants that have claimed part of the street as outdoor cafes, with its boutiques and small specialty stores, has become the Champs Élysée of the Western Hemisphere. And Broadway—with its serried ranks of shops, block after block, mile after mile—is almost impossible to explore thoroughly. Every day you walk along it you are bound to discover something fascinating and new that you have overlooked before or that has come newly into existence.

But New Yorkers? Are they not rude and gruff and overbearing?

No, they are *not*. That is the mythology again.

They may seem so, but we are dealing here with the exigencies of life in a densely packed island (Manhattan) inhabited by people with a keen sense of what will make life bearable.

Imagine life in a small town where you know everyone and everyone knows you. If you walk down the street, you must greet everyone you pass: "Hello, Mrs. Miller. How are you? Has your husband recovered from his cold? How does your daughter like married life? You're looking very well today."

To refrain from doing this is to insult a neighbor, but how wearisome it must be to be forced into small talk at every meeting, and how uncomfortable to know that your every moment is marked down and every facet of your life is known and commented on.

This sort of thing would be impossible in New York. Living in such close quarters (there are over two hundred families in my apartment house, all constantly using the elevators and the lobby as we move in and out), small talk would kill us all within a week. With the streets consisting of solid masses of humanity moving in both directions, for all to smile on each other as they pass would quickly paralyze every face in the city.

The only way we can retain our privacy is to ignore one another and remain unaware of each other's existence.

And that is so good a thing I would not exchange it for all the synthetic friendliness and make-believe amiability in twenty million tons of greeting cards. In New York I can be as eccentric as I please. I can wear what clothes I wish, I can walk with my hands in my pocket and a whistle on my lips, I can skip if I am happy, frown and scowl if I am annoyed—no one will care, no one will pay attention to me. I can be me in any fashion I wish and you can be you, and all of New York will grant us the permission of indifference. You can get that nowhere else in the world, I believe, and once you've grown to appreciate the freedom that indifference grants you, you will never want to give it up.

None of that, however, makes a New Yorker less than human. Be in trouble, ask for directions or help of any kind, and the walls of indifference will tumble.

Newspapers are full of letters from tourists who have been amazed to discover that New Yorkers in emergencies are kindly and decent.

A year ago, my daughter, Robyn, moved to New York and now lives, with her two cats, just two miles southwest of me. She had lived all her life in Boston, and I was nervous at how (or whether) she would acclimate herself to the Big Apple. When she heard I was going to write this article, however, she gave me strict instructions to tell the world that she also loves New York and wouldn't live anywhere else for any reason she can currently imagine. And my dear wife, Janet, agrees—which makes it unanimous.

34

THE IMMORTAL
SHERLOCK HOLMES

When Arthur Conan Doyle was twenty-six years old, he was showing no signs of success. He was a physician, but his medical practice was bringing in very little money. For some years he had been writing an occasional short story and that, too, brought in very little money. It is not surprising, then, that he was in debt and that his creditors were growing impatient.

In March of 1886, however, it occurred to him to write a detective story, and to model his detective on Dr. Joseph Bell, who had been one of Doyle's teachers in medical school. Bell was an observant man, with a tremendous memory for minutiae. From small indications he could diagnose problems before the patient had said a word. He could even give his patients, almost at once, details concerning their personalities and private lives.

So Doyle wrote *A Study in Scarlet*. In it appeared the detective, Mr. Sherlock Holmes, and his faithful companion, Dr. John Watson, to say nothing of a pair of Scotland Yard bunglers.

At first the new story, the length of a short novel, seemed no more successful than Doyle's previous literary efforts. It was rejected several times and was finally bought for the flat sum of twenty-five pounds ($125—which was, of course, worth a great deal more then than it would be now). It appeared about December 1, 1887.

It did not take the world by storm, but there was interest in another Sherlock Holmes story, so Doyle wrote a second short novel, *The Sign of the Four*—taking time out from a historical novel to do so. That did not take the world by storm either, and Doyle decided to write a series of short stories involving his detective.

The first of these, "A Scandal in Bohemia," appeared in *Strand* magazine in July 1891, and at last there was an explosion. With each further story the popularity of Sherlock Holmes increased, and within two years Sherlock Holmes was the most popular fictional character in the English-speaking world.

Doyle was not pleased. He was rapidly growing rich, but he was also receding into the shadows. No one knew *him;* they knew only Sherlock Holmes. What's more, no one was interested in his other writings, which he felt were far superior to the Sherlock Holmes stories; the public wanted only the detective.

In 1893, therefore, Doyle wrote "The Final Problem," in which he simply killed Sherlock Holmes off and, he hoped, got rid of him forever.

If that is what Doyle thought, he proved monumentally wrong. The vast pressure of an enraged public forced him to write a new Sherlock Holmes novel, *The Hound of the Baskervilles,* in 1901. Then in 1903 he surrendered completely and wrote *The Adventure of the Empty House,* in which he brought Sherlock Holmes back to life. He then continued with other short stories, so that there were, in the end, fifty-six Sherlock Holmes short stories and four novels. The last Sherlock Holmes story was published in 1927.

In later life, Doyle grew interested in spiritualism and was rather pathetically nonrational in this respect. I often think that this was, at least in part, an attempt to appear to the public in *some* form other than as the creator of Sherlock Holmes.

In 1930 Arthur Conan Doyle died, at the age of seventy-one.

Doyle was right. Sherlock Holmes was a monster who destroyed his creator. Even in his lifetime there were satires and imitations of the great Sherlock, and these continued without pause after Doyle's death. Doyle, obscured by the detective in his lifetime, sank almost to nothing after his death, while Sherlock continued—untouched and immortal.

Although Sherlock Holmes had his predecessors (notably Edgar Allan Poe's detective, C. Auguste Dupin), it was Sherlock who took the world by storm and who changed the nature of the detective story. For the first time the genre took on an adult, literary character.

What's more, almost every succeeding detective story writer paid homage to Doyle's great creation. Sherlock Holmes is mentioned frequently in detective stories as the epitome in detection. I have characters in my own stories refer to my own detective, Henry, as "the Sherlock Holmes of the Black Widowers." There is simply no other way of describing a clever detective.

Once the copyright on the Sherlock Holmes stories ended and they entered the public domain, writers at once began to write additional such stories without any attempt at masking their efforts. Sherlock Holmes simply came back to literary life once again, as did the ever-faithful Watson.

Nor is it only in print (and in every language, for Sherlock's popularity is by no means confined to English-speaking individuals). No other fictional character has appeared so often on the stage, on the screen, or in television, as Sherlock Holmes.

And Sherlock's life is not confined to fiction, either. So careful was Doyle to give all sorts of casual details concerning Sherlock's appearance, personality, and lifestyle, that the less sophisticated readers took it for granted that Sherlock Holmes was really and literally a living detective who dwelt at 221B Baker Street. Letters poured into London's post office from all over the world, as problems great and small were presented to the remarkable detective. No other fictional character has ever been so literally alive to so many.

Nor is it only the unsophisticated who think Sherlock Holmes is alive.

There are highly literate and educated human beings who enjoy the books and the character so much that they *prefer* to believe he is alive. This may be only pretense, but it is considered bad form to admit that.

In 1934, four years after Doyle's death, a group of Sherlock Holmes devotees founded "The Baker Street Irregulars" in the United States. (The original Irregulars were a group of bright neighborhood youngsters used by Sherlock Holmes to help him gather information.)

The Irregulars have flourished ever since and, in fact, on January 6, 1984, celebrated their semicentennial.

The Irregulars have made Doyle's worst fears come true. To every true Irregular, Sherlock Holmes was a living person (and still is, for no Irregular will admit that Holmes has ever died) and John Watson is the true author of the stories. While Arthur Conan Doyle's existence is grudgingly admitted, he is virtually never mentioned, and when he is, it is only as Watson's "agent."

The great game of the Irregulars is the thoroughgoing analysis of the "sacred writings." One might determine, from internal evidence, exactly when this adventure or that took place. One might try to determine, from internal evidence, the number of different women to whom Watson was married, or whether Holmes went to Oxford or Cambridge.

Since Doyle wrote the stories hurriedly, and after the first dozen or so with deep distaste, there are numerous contradictions in them—but these are never to be accepted as contradictions. Some theory must be advanced to account for the *appearance* of contradiction. (Thus, when Mrs. Watson refers to her husband as James, rather than John, this must not be considered a slip of the pen. The accepted explanation is that Watson's middle initial, H, stands for Hamish, which is, of course, the Scottish version of James.)

How, then, can we account for the great and continuing popularity of Sherlock Holmes? It is nearly a hundred years since his first appearance and yet his colors have not faded. We may be embarrassed by his use of cocaine (at a time when its addictive properties were not yet understood), and wonder what he must have smelled like in view of his incessant smoking of strong tobacco, but we forgive him that, and all other faults, because we love him so.

But why do we love him so?

Well, if we think about it, it must seem to us that there are only a couple of classes of fictional heroes in the world.

First there is the hero who is renowned for his muscles and physical prowess. Almost every culture has its muscular superman. The Sumerians had Gilgamesh, the Greeks had Heracles (Hercules), the Hebrews had Samson, the Scandinavians had Sigurd (Siegfried), the Persians had Rustem, the Irish had Cuchulain, the medieval Europeans had Lancelot and Roland, and so on. Even contemporary writings have their Tarzans, their Conans, their Doc Savages, and so on.

These heroes appeal to us strongly because their solutions to problems and their escapes from danger are so simple. They just bash their enemies on the head, and that's it. Which one of us has not, every now and then, longed for the ability to do just that to some bully or petty tyrant?

With muscles that are large enough, you don't have to be smart. Even when muscle-men are presented as intelligent—as in the case of Lancelot or Doc Savage—it is not their brains but their brawn that gets them out of trouble. Sometimes their lack of intelligence is emphasized as an actual virtue—they are just simple folk who attack problems without the need for sophisticated subtlety. Or as Anna Russell describes Siegfried, the hero of Richard Wagner's *Der Ring des Nibelungen,* such heroes are "very brave, very strong, very handsome, and very, *very* stupid."

Of course, the possibility then arises of a second class of heroes, those who use their *brains* to overcome problems. Opposed to the Hebrew Samson is the young David, who, facing the huge Goliath, uses his sling and attacks the giant from a distance. Opposed to the Greek Hercules is the Greek Odysseus (Ulysses), who faces problems with unfailing shrewdness.

Somehow, though, the clever man slips away from true heroism in the imagination of the audience. Perhaps the ordinary member of the audience can visualize himself brawny and muscular, but can't visualize himself shrewd and brainy. As a result the clever man can only attain a limited popularity.

The Davids of the world are not really brave; they fight from a distance; they don't face the enemy "like a man." And the Odysseuses of the world are shrewd and deceitful; they are clever liars, smooth tricksters. They don't win in straightforward fashion, but through stratagems that catch the simple, strong men unawares.

In fact, just as the strong man's stupidity is made into a virtue, the shrewd man's brains become a vice. Cleverness taken to an extreme makes a man able to manipulate not only his enemies but the Universe. The clever man has learned the charms, the spells, the magic that can bend supernatural forces to his will. He has become a "wizard," or a "sorcerer."

There are, of course, wizards and sorcerers who are on the side of good. There is the Welsh Merlin, and the Finnish Vainamoinen, but these are definitely in the minority. For the most part, the stereotype is of the "wicked sorcerer" and, doubly so when the villain is female, "the wicked witch."

In the Arthurian legends, opposed to the good Merlin, there is the wicked witch, Morgan le Fay, a much more vivid and memorable character. Morgan surely drowns out Merlin, just as in the movie version of *The Wizard of Oz,* the Wicked Witch of the West totally erases the vapid Good Witch, Glinda.

To the average man, there is danger in intelligence. The strong man, being stupid, can be bent to one's will, even to the will of the average man, for the strong man is *very* stupid. The clever man, however, cannot be relied on. He can always turn on you for reasons of his own.

For that reason almost all simple tales of good and evil have good represented by the stupid, strong man, the slow thinker with the fast gun, the childlike overgrown innocents like Li'l Abner and Joe Palooka. The evil is presented by the shrewd and cunning crooks, the evil scientists, the scheming lawyers and greedy bankers.

In fact, just to put everything flatly on the table, there is a very well-known form of pop literature just now called "swords and sorcery." Swords represent good and ignorance, while sorcery represents evil and intelligence.

Where, in all this, does Sherlock Holmes fit in?

Doyle has discovered a way to disinfect intelligence and make it tolerable and nonfrightening to the general reader.

Sherlock Holmes is the epitome of intelligence, and he is always on the side of good. He defeats the villains by his superior brains. But he avoids becoming a villain because he refuses to be obsessed by shrewdness and trickery. He *is* shrewd, however, and he is not above using trickery, as in donning a clever disguise. These mental abilities, however, are not his chief weapons.

His chief weapons are acute observation and keen logical deduction.

The sorcerer never reveals his tricks to the audience. He can control the Universe and turn his opponents into frogs by spells that are not given, or that, if given, are clearly nonsense, since when *you* repeat them, nothing happens. In the presence of such unrevealed tricks, the audience can only be afraid. Unexplained powers are inimitable and therefore dangerous.

Sherlock Holmes, however, though his tricks seem equally mysterious, is always ready to reveal them as "elementary, my dear Watson." Before the end of the story, he is sure to reveal his line of reasoning so that all seems plain and clear. You could have done it yourself. Sometimes, outside the plot of the tale, he does a bit of legerdemain, deducing much about Watson's brother, for instance, by merely examining a watch that had once belonged to him. Watson is thunderstruck until Holmes explains how he does it, and then it *does* seem elementary.

The audience meeting such a unique character cannot help but be delighted. How wonderful to have intelligence made so plain and straightforward that any reader can feel he might be that intelligent too, with a little bit of luck. (Watson assists us here. Presented as brave and loyal, he is so dense that almost any reader feels smart by comparison.)

Sherlock Holmes is swept into our arms, therefore, and—especially since Doyle writes with such pleasant simplicity and clarity, and is so careful to make both Holmes and Watson three-dimensional and believable—we cannot bear to believe he is not alive.

Surely, there have been thousands of Sherlock Holmes's imitations since 1887, but it was Sherlock Holmes who has had the headstart, and no one, starting later in the race, can catch up.

I am trapped by Sherlock, too. Thinking about it coolly, I come to the clear conclusion that Agatha Christie was a better mystery writer than Doyle was, and I think that Hercule Poirot is a cleverer fictional detective than Holmes is—but even to me, Sherlock Holmes is more "real," more "lovable." Even his defects, his cocaine and tobacco, his inability to love and to let himself be loved,

his occasional cruelty to Watson, make him all the more "real."
So I, too, am a member of the Baker Street Irregulars.

35

GILBERT & SULLIVAN

Some things simply seem to go together: ham and eggs, corned beef and cabbage, hot dogs and mustard, lox and cream cheese on a bagel. Such things can be eaten separately or in other combinations, but they never have quite the satisfactory taste that they have together.

It's sometimes that way with fictional couples. Can you imagine Romeo without Juliet, or Antony without Cleopatra, or, for that matter, Rhett without Scarlett?

And, in one case I can think of, it works for a pair of real-life people, where the ampersand in between the names is far more essential than either name.

William Schwenk Gilbert was born in 1836, and throughout his life he was a cantankerous curmudgeon who worked very hard to earn social unpopularity and invariably succeeded in this endeavor. He was a most unlikely person to decide to make a living as a humorist, but he succeeded in this endeavor also—after a fashion.

He wasn't a humorist to start with, of course. He tried the army and didn't enjoy it, he tried being a bureaucrat and hated it, he tried being a lawyer but had no talent for it. However, midcentury was the time of an efflorescence of humor magazines in Great Britain, and Gilbert attempted to earn a few shillings writing funny pieces for them.

Gilbert succeeded at this, particularly since he showed an unparalleled talent for writing comic verse, to which he appended delightful comic illustrations. This verse was eventually collected as *Bab Ballads* ("Bab" being used by Gilbert as a signature for his illustrations).

Gilbert's humor was of the type we would call "dark" (sometimes very dark indeed). The best Bab ballad is "Etiquette," in which the humor arises out of two men being marooned on an island after a shipwreck and in their not being rescued. The most frequently reprinted Bab ballad is "The Yarn of Nancy Bell," which tells the story of another shipwreck; cannibalism is thrown in for additional laughs.

Nevertheless, comic verse has no great lasting value, and except for an occasional appearance of a Bab ballad in poetry anthologies of the lighter sort, Gilbert's ballads would probably be forgotten today if they were all we had.

Gilbert also began to write plays, some of which were simply comedies (occasionally in blank verse) and some of which contained verses that could be set to music by some nonentity or other. He wrote about seventy plays alto-gether that have by now completely vanished from world consciousness, al-though they were moderately popular in their own time.

To be brief (and cruel), Gilbert, as Gilbert, left no lasting mark, and there is no reason on Earth why I ought to be writing about him now; that is, about Gilbert, *as Gilbert*.

Arthur Seymour Sullivan was born in 1842 and even as a teen-ager showed unmistakable musical talent. His first published hymn appeared when he was thirteen years old. A year later he won a coveted musical scholarship, and by the time he was twenty he was acknowledged as Great Britain's foremost composer. (To be sure, Great Britain was passing through a period then in which it was a musical desert.)

Sullivan was amazingly successful in other ways, too. He was convivial and sociable. He was a snob who sought out the company of the upper classes and was accepted by them. To onlookers he must have seemed like a golden boy, even though he was troubled, as life went on, by ill health and reverses in his business dealings.

Sullivan turned out a grand opera, and innumerable oratorios, hymns, songs, symphonies, odes, and so on and so on, all, or almost all, to con-temporary acclaim; and almost all are now totally forgotten and are played, if they are played at all, only as curiosities. (There are two exceptions. Sullivan wrote the music to the still well-known "Onward, Christian Soldiers" and "The Lost Chord.")

Therefore, Sullivan, as Sullivan, left no lasting mark, and there is no reason on Earth why I ought to be writing about him now; that is, about Sullivan *as Sullivan*.

In 1896 Gilbert and Sullivan met for the first time, and Sullivan took an instant dislike to Gilbert (something that was easy to do). Sullivan therefore agreed only reluctantly, in 1870, to write the music for a comic opera, "Thespis," which Gilbert had written. It opened in 1871 and it was a flat failure. (The music to it was never published and no longer exists.) Gilbert returned to his play-writing and Sullivan to his music, each continuing to score contemporary success that would not survive their lifetimes.

But then in 1874 a theatrical manager, Richard D'Oyly Carte, was planning to stage *La Perichole,* a very successful French opera by the foremost light-comedy composer of the day, Jacques Offenbach. *La Perichole* was not quite long enough, however, and D'Oyly Carte wanted a curtain-raiser to round out the evening.

Gilbert had just written a short all-verse satire on the legal profession called *Trial by Jury,* inspired by one of his own Bab ballads, and D'Oyly Carte liked it. He persuaded Gilbert to read it to Sullivan, who also liked it, and who agreed to write the music.

On March 25, 1872, *La Perichole* opened, with *Trial by Jury* thrown to the audience as simply a piece of fluff designed to get a few smiles before the real fun of the evening began.

What followed was unbelievable. The audience went mad, not over *La Perichole,* but over *Trial by Jury.* They continued to go mad, night after night, week after week, month after month.

What had happened was that something new had been born, "Gilbert & Sullivan." Gilbert was just Gilbert, Sullivan was just Sullivan, but "Gilbert & Sullivan," with the ampersand, were an immortal pair that, nowadays, over a century later, are as popular as they ever were. What's more, there seems no danger, to discerning eyes and ears, that they will ever cease being popular.

Of course, *Trial by Jury* might have been an aberration, a one-shot. The next Gilbert & Sullivan opera, *The Sorcerer,* was successful, but it did not inspire the madness *Trial by Jury* had. But then, on May 25, 1878, the Gilbert & Sullivan opera *H.M.S. Pinafore* opened, and the public went mad again. The new opera became nothing less than a national craze. It jumped the ocean and was pirated by American producers (there was no international copyright law then) and met with equal success there.

Next followed *The Pirates of Penzance, Patience, Iolanthe, Princess Ida, The Mikado* (most successful of all), *Ruddigore, Yeomen of the Guard,* and *The Gondoliers.* These eleven Gilbert & Sullivan plays remain alive and are constantly being revived in all fashions from high-school efforts to major productions in the theater and on the screen.

Unfortunately, however successful Gilbert & Sullivan might be as a team, the component individuals disliked each other with increasing venom. Sullivan felt he was slumming when he wrote musical comedy and his snobbish friends urged him constantly to turn to "great" music. As for Gilbert, he was pathologically alert to fancied slights, certain that D'Oyly Carte was siphoning off some of the profits, and indignant at his constant feeling that Sullivan was trying to make his music take precedence over the words. In 1890, soon after *The Gondoliers,* a brilliant success, opened, Gilbert and Sullivan quarreled over a minor expenditure, and the partnership ended.

Gilbert continued to write plays and Sullivan to write music, including an opera *Ivanhoe.* Once again, everything was forgettable. Even when, in desperation, they reunited and did two last musical comedies, *Utopia Limited* and *The Grand Duke,* the magic was gone. Those two are rarely performed.

Oddly enough, Gilbert's best Bab ballad, "Etiquette," written before he met Sullivan, sounds like a prediction of the partnership. In the ballad, two shipwrecked Englishmen, ignoring each other on their island refuge, live miserably and unhappily. They become acquainted and there follows a happy time. But then they quarrel over something totally insignificant, and there is a return to misery and unhappiness. And so it was, in its own way, with Gilbert and with Sullivan.

But what is there about Gilbert & Sullivan that makes them immortal?

Naturally, I can't speak for others, but what is it that makes *me* so crazy about them? I have seen and heard each play innumerable times (even *Utopia Limited* and, once, *The Grand Duke*). I know every word and every note, and yet I rarely miss any performance I can reach. Though no word and no note catches me by surprise, every word, every note delights me. Why?

In the first place, it's a perfect fusion. Words and music are equally witty, and no one would want to give up either. There is no place in Gilbert & Sullivan (as there is in all other musical comedies) where we find a piece of music delightful but are irritated by the sapless verses that have been matched to it, or where we admire a clever set of lines but find them cursed with uninspired music.

In fact, it is fun to argue over which is more important to the plays, Gilbert or Sullivan; it is fun precisely because there is no answer. One can argue either side.

For instance, for every straight performance of Gilbert & Sullivan, there must be at least ten parodies. There is no amateur versifier who hasn't attempted to produce some homemade musical farce for a company picnic or for a school play or for some other trivial occasion, and the choice is, nine times out of ten, to parody Gilbert & Sullivan. I have myself participated in such things on a number of occasions.

In such parodies, Gilbert's words must, of course, be altered. Sullivan's music, however, remains totally untouched. Not one note is changed. With altered (and incredibly inferior) words and Sullivan's music, the parody remains Gilbert & Sullivan, and, however amateurish, retains a distant glow of the grand original. Doesn't this prove the superiority of Sullivan over Gilbert?

Ah, but wait. There are many operettas in which the music is delightful. Composers such as Offenbach, Strauss, Friml, Romberg, Rodgers, Lowe, needn't take a back seat to Sullivan. And yet how much do people know of the words to the songs these others composers have written? There's a great song that starts, "Come, come, my heart is calling you. . . ." You can probably hum the rest of the tune, but what are the rest of the words? What are the words to the song in which one of the lines is "Lover, come back to me"? For that matter, how does "Oh, what a beautiful morning" start?

I'm sure that a number of you can answer all these questions, but not many. Gilbert & Sullivan enthusiasts, however, don't just go around humming the tunes. They *know the words*. That is because the words are *clever*.

Gilbert uses the full range of the English language in his verses. He rhymes "diminutioner," "ablutioner" and "*you* shun her." He rhymes "mathematical" and "quadratical," "cuneiform" and "uniform," "opportunity" and "impunity," "felicity" and "domesticity," and so on. No "moon," "June," "spoon" for him.

No one before Gilbert wrote verses as he did, and since his time only Cole Porter can be mentioned as a rival in witty versifying, and Porter admitted to being an admirer and student of Gilbert. (On the other hand, Porter wrote his own music, and grade-A music it was, too.)

What's more, Gilbert's book for each musical was sharp and biting satire that continues to work even a century later. He is perhaps a little too fond of puns for modern tastes, but, that aside, he still hits his targets with savage success.

Finally, Gilbert directed and stage-managed his plays and insisted (and, oh boy, when Gilbert *insisted* there was no backtalk, you can bet) that every word of every song be clearly heard over the music.

What's more, there's the fun that adheres to that part of the play that is neither words nor music. Gilbert & Sullivan plays lend themselves to overacting. It is impossible to go too far in this respect, and since every actor on the stage loves to go all out, the audience has the opportunity of watching all the characters having a good time, and that good time is catching. Gilbert also insisted that the chorus play an active role throughout, which adds another element of fun to the play.

In fact, each particular production of the play can exert its full creativity on the set, the costumes, and the "business." When, like me, someone in the audience knows every word and every note, he or she can still be delighted and surprised by the way those words and notes are delivered and by the choreographic interplay of the anywhere up to forty characters who may crowd the stage.

The words and notes may be the same, but I have never seen two Gilbert & Sullivan productions of any particular play that were exactly alike and that weren't each in its own way (with very few exceptions) delightful.

36

MENSA AND I

I first heard of Mensa in 1961, when a charming young woman, Gloria Salzberg, who was herself a member, urged me to join. I was quite willing until I heard I would have to take an intelligence test in order to see if I could qualify.

With the courage of an Asimov, I promptly quailed. I had not taken an intelligence test for many years, and I didn't think I any longer remembered the answers. I therefore tried to get out of it.

"Isn't my life, my profession, my books, the obvious evidence of my intelligence?" I demanded.

"No," said Gloria sweetly. "Here is the test. Here is a pencil. I will supervise."

She stationed her husband (a large, muscular specimen) at the door to prevent my escape, and I took the test. Fortunately (and I don't suppose it would happen again in a hundred years) I passed and was made a member.

For several years, I attended meetings with vigorous assiduity. (That's just Mensa talk. I'm not sure what it means.) Then I stopped.

You see, I am what is called a prolific writer. I have had 355 books published (so far), and hundreds of short stories, and thousands of essays, and you may think these things all write themselves, but they don't. Just *copying* all that stuff would take me nine hours a day, seven days a week. Making them up in the first place takes me an additional hour every single day.

It all earns me a moderate living, but the other side of the coin is that I have no time for anything else—so I just quietly dropped my Mensa meetings and, being a frugal person, stopped paying my dues.

As time passed, I moved to New York, and in 1972, Victor Serebriakoff, the Ayatollah of Mensa, visited New York and demanded to meet me for some nefarious reason of his own. (He's never had an un-nefarious reason in his life.) I was not proof against the imperious summons. There I was staring at this five-foot-five fellow with a seven-foot-seven charisma.

"Why," he demanded, "have you allowed your Mensa membership to lapse?"

I tried to explain.

He dismissed the explanation with an impatient "Tchah!" (which may be Russian for "In your hat," but I'm not sure). Then he said, "Just renew your membership."

I demurred (more Mensa talk).

"You might as well," he said with a Serebriakoffian snarl, "because if you don't, I'll pay your dues for you and we'll list you as a member anyway."

With the typical pride of an Asimov, I said, "Okay," but my dear wife, Janet, wouldn't let me, and I joined up once more at my own expense.

Victor then revealed the basic scheme that lay behind all this. He began a shrewd campaign to get me to come to Great Britain, even though all my instincts are against travel. He seized me by the throat and said, "You will come or I will strangle you."

I was unable to resist the subtlety of the approach, so in 1974 I made the trip. There I discovered the full depth of Mensa hospitality, the beauty of the English countryside, the charm of London, and most of all, *most* of all, the absolute excellence of Mensa people as an audience. (It probably has to do with their intelligence, now that I come to think of it. After all, I'm one of them, so it's hard to hide those things from me.)

And as I was about to leave, Victor said to me, "You are one of the two International Vice-Presidents of Mensa."

With the typical penetration of an Asimov, I said, "What?"

He repeated his remark, and I said I didn't know how to run for the post and that I didn't think the Mensans would vote for me.

"That's all right," said Victor with Serebriakoffian hauteur (we Mensa people speak French), "this is a democratic organization and I've just appointed you."

I'm afraid to ask, actually, but I think I'm still International Vice-President.

37

WRITE, WRITE, WRITE

For about three decades, I have considered myself, and been considered by others, to be a prolific writer. However, in my correspondence with a gentleman named James Fogarty, he referred to me as a "chronic writer."

I was amused. It might appear, somehow, a better term, for "prolific" seems to refer to little more than the quantity of written material, whereas "chronic" implies an abnormality, almost a sickness. A prolific writer may just be working hard, but a chronic writer (you might judge) can't help himself because he is in the grip of a vitamin deficiency or a hormonal imbalance.

Certainly that is the feeling I often get from those who question me on the matter. They ask, curiously, "How do you do it?" as though they feel it to be a peculiar trick, like that of someone who makes lighted candles appear between his fingers. Or they say, doubtfully, "You must have a great deal of discipline to keep to the typewriter so steadily," as though I am doing something incredibly distasteful through an astonishing exercise of will. And sometimes they ask, pityingly, "Do you ever regret having to do all the work you do?" as though I am peering longingly, through bars built of typewriter keys, at the great wonderful world outside.

Every once in a while, I am moved to try to explain the situation (as far as I can) and to answer the questions I am asked. It strikes me that it may even be useful to do so, since it just may be that there are people out there who would like to learn how to be prolific. There are, after all, advantages to it.

If you are a successful writer (that is, one who sells what he writes most of the time) and if you can teach yourself to be prolific as well, then you will have that much more to sell and you will probably earn a larger income. Then, too, prolificity attracts attention, catches the eyes of editors, and brings you more assignments without your having to beat the bushes for them or undergo the indignity of writing humble letters of inquiry. You may even gain a certain fame in nonliterary circles through the sheer quantity of your output, so that your name becomes that much more generally recognizable and any books you write may sell that many more copies to the benefit of your bank account.

Well, then, how do you go about it? Let's list some of the rules that I have worked out by thinking about the matter, now and then, over the decades.

1. *You can't start from scratch.*

If you have no experience in writing, you can't say to yourself, "I think I'm going to be a prolific writer" and sit down to write with that in mind.

First, you have to be sure you can write. That requires a certain innate ability (or talent, if you prefer) to begin with. It requires a certain amount of education, which will give you a decent vocabulary and an understanding of the mechanics of the language, including its spelling and grammar. It requires a certain amount of intelligence and imagination and a sense of application and perseverance which will keep you reading the works of good writers, observing their techniques, and surviving the difficult period in which you gradually learn how to do it yourself.

(No one else is going to teach you, by the way. Others may direct you to proper reading, point out elementary mistakes, and give you a general rule or two, but these are trivialities. In the end, you will have to learn for yourself the necessary subtleties of writing. After all, no amount of urging by bystanders to "keep your balance" will help you learn to ride a bicycle till you get the feel of keeping it upright by yourself.)

But then, supposing you *are* a writer, where does the prolificity come in?

2. *You have to like to write.*

You may like being a writer but that's not the same thing. Indeed, many people who are not writers at all get ideas of subjects concerning which they would like to write, and they can daydream pleasantly of writing it. It must seem delightful to them to be a writer.

And almost anybody, it seems to me, would be overjoyed to actually have a book in his hand that he has written. It would then *surely* seem delightful to be a writer.

But what about the in-between? What about the part that comes after the thought of what you want to write and the product that you have written? What about the actual matter of thinking up words and of writing them down one after the other and adjusting them one by one till they are right?

That is a tedious and uncertain process that seems to have little to recommend it. Even good writers, even *driven* writers, rarely like the process. That is why writers are so commonly pictured as moody, bad-tempered, or even prone to drink. Why not? They are assumed to be sensitive souls, forced (either by the necessity of earning a living, or by their own impulses) to do something they find very unpleasant.

And if it should turn out that you simply don't like the mechanical process of writing, then you cannot be a prolific writer. You can be a *writer,* of course, and a *good* writer, too, even a *great* writer, but you will only be able to force so much out of yourself, and even that with considerable distress.

Clearly, it is sufficient to be a good writer, and one might easily argue that it is better to be a good writer with little production than a poor writer who turns it out in reams. It is that thought that must console you if you hate to write and yet bemoan your failure to be prolific.

Suppose, though, that you hate to write and yet find no satisfaction either in moaning or in shooting for quality, but are determined to be prolific. Is there any way you can learn to love the production of torrents of words?

The problem, here, may be either with yourself or with the process. If you consider the fault to lie with yourself, then you may try to deal with it by hypnosis, or psychotherapy, or with any of the flood of self-help books that are published each year. Concerning that, I have nothing to say, since such devices are highly individual and what works for one may not work for another.

If, on the other hand, you are convinced that the fault lies with the general difficulty of the process, then it might be fair to try to simplify it, and there at least we strike something that may be general. If a process is simplified for one person, it is very likely to be simplified for others, even for most people.

3. *Be spare.*

Styles vary, from the extremely ornate to the extremely direct. It seems to me that the more ornate the style, the more difficult it is to write it well, and the greater the number of possible alternatives one must face. Inevitably, you find uncertainties besetting you. Is this sentence written as effectively as it might be? Is this a beautifully turned paragraph or simply hogwash? And how can you tell?

A good writer *can* tell and he *can* arrange to turn out effective material couched in poetry and metaphor, but even he (or she) must be slow and thoughtful about it and must be content to revise frequently in order to try different routes to his destination and determine which is best. It takes time, a lot of time. It may be that Shakespeare turned out incredibly beautiful material as fast as he could write, but there was only one Shakespeare in all of history.

On the other hand, you can be spare. You can eschew poetry and metaphor and say what you want to say as directly as you can possibly manage. That is my own system. I never use a long word if a short one will do; or a complex sentence if a simple one will get the thought across. I don't bother with figures of speech or literary allusions if I can manage to do without. What is more, I keep all my action (when I write fiction) on a bare stage. That is, the only characters in a scene are those who are actively doing or saying something, and I describe neither those characters nor the background to any greater extent than is required to make matters clear to the reader.

With matters simplified in this way, there are fewer alternatives possible. You can then be reasonably confident that what you have written is the rational way of saying what you have to say, and you inevitably decrease your worries concerning it. With enough practice, you stop worrying altogether and you become self-assured. Once that feeling that you are writing junk leaves the pit of your stomach, most of the unpleasantness associated with writing is gone. What's more, you can write more rapidly and are on the way toward prolificity, if that is what you want.

4. *Be a juggler.*

Even if you like the mechanics of writing and if you write rapidly, your problems are not over.

Lying in wait for all writers is that horrible and fearsome monster: the writer's block.

Very few, it seems, are immune. All writers dread that moment when one is staring at a sheet of paper and nothing comes. If, after a long time, one squeezes out a sentence, it seems to lead nowhere, and there is nothing to do but rip out the sheet of paper and insert another.

(There you have another popular conception of the writer—unshaven, haggard, staring blearily at a blank sheet in his typewriter, ashtrays overflowing with cigarette butts, coffeepot empty, the floor about him strewn with crumpled sheets of paper.)

I have known writers—*good* writers—who have been victimized by writer's block for many years—nay, decades—and who can't manage to get the juices flowing even when driven by want and penury.

Clearly, a prolific writer can't remain prolific if he is stricken with writer's block very often or for very long. In my forty-eight years as a professional writer, I have never been stricken by the disease. I have avoided it by deliberately multiplying my commitments.

After all, any particular piece of writing can pall on you. Your well of invention may run dry, or you may just sicken of your characters and their problems. It might then be wise to put the thing aside and let it ripen a while until you return to it with new ideas and a fresh approach.

The trouble is that while you are doing nothing, the fear may grow that you will never be able to do anything again. The longer you stay away, the greater the fear may be, and that fear may itself produce the actuality. And as soon as you experience the actual block, that experience itself seems to strengthen the block in a vicious cycle that seems to have no end.

If, however, you are forced, for one reason or another, to abandon a particular writing project, why does that mean you must stop writing? Suppose you have *two* projects on hand. In that case, if one is abandoned temporarily, turn to the second. Since the second will represent fresh challenges and difficulties, you will find it stimulating, and the mere fact that you are writing steadily and productively means that no thought of a block need enter your mind. Eventually, the second project may pall, or new ideas concerning the first may have entered your head, and you need then merely turn back to the original job.

In my own case, I have for decades kept busy with three or four different books plus a number of smaller pieces, so that every morning I look at the list of jobs I have in progress and can choose to work on the one that seems most interesting at the moment. And if in midflight I become anxious or bored, I can simply shift to any of the others.

Nor do I necessarily wait for something to force me from one to another. I become a juggler, switching from one task to another, just for the pleasure of it, and thus I keep writer's block successfully at bay.

5. *Be businesslike.*

Can we juggle forever? Writing is not a game and deadlines exist, especially for short pieces. If a piece must be finished by tomorrow, then you can't decide to do something else just because you feel like it. You have to do this particular job, and do it by tomorrow.

You must then sit down and do it, even if you would rather not. That happens every once in a while, but so what? There are always times when something must be done, willy-nilly, in any job, and there's no use treating writing as something ethereal and "artistic." A prolific writer has no room for that sort of silliness.

This is especially true where there is a question of deciding what to write in the first place. That is, perhaps, the most common question of those asked any prolific writer. "Where do you get your ideas?" (I have a writer friend who, whenever asked that question, answers, "From Schenectady. There's an idea factory there to which I subscribe and every month they send me a new idea." I think some people go away believing him.)

Most people who ask that question seem to think that ideas come from "inspiration," and that this is encouraged by some rare and arcane form of stimulation—the careful recital of charms or the use of esoteric forms of witchcraft.

But that's nonsense, of course. No prolific writer can wait for "inspiration." No doubt, there are thoughts that sometimes well upward from some hidden spring of the unconscious, originated by some casual remark or some chance sight—but you can't rely on it.

What *can* you rely on?

In the first place, one can depend on variety. I write two short pieces a week on the average. Some of them are fiction, some are nonfiction. The fiction might be science fiction, fantasy, or mystery. The nonfiction might deal with science, with the future, with some matter of opinion, or with none of the above. In any case, there is no need to be forever coming up with the same class of idea, and that helps.

There are, to be sure, prolific writers who cling to very much the same sort of thing year in and year out. In fact, there are writers more prolific than I am who turn out nothing but mysteries of a particular type, or romances, or juveniles. My own feeling, though, is that this involves the danger of having one's brain turn to mush, and of falling into the trap of endless and stale repetition.

Variety will keep you fresh and make the task comparatively easy.

You must rely also on the world about you. You must look and listen sharply, for almost anything can serve as the nub of an idea. At dinner a friend of mine told me of his glasses, which darkened on exposure to the sun and lightened once more in the shadow. He tried to make his wife feel guilty by pretending he had been waiting an hour for her in the lobby of a hotel when actually he had just arrived himself. She pointed out that his glasses were still dark and instantly I had an idea. A casual acquaintance once told me that his bank's computer system broke down as soon as he entered the place to inquire as to his bank balance and *that* gave me an idea. Any item in a newspaper, any scene in a television drama, may serve.

But what if everything fails? If you don't pick up any idea from the environment about you, what is there to fall back upon?

Upon yourself. The most common and the most effective method for getting an idea is to sit down and think as hard as you can. This is not the case for writers only. Edison said, "Genius is two percent inspiration and ninety-eight percent perspiration." When Newton was asked where he managed to get his remarkable insights, he said, "By thinking and thinking and thinking about it."

This is disappointing news, I suppose. When people ask me where I get my ideas, I always say, "By thinking very hard," and they are clearly disillusioned. Anyone, they seem to decide, can do *that*. Well, if anyone can, then *you* do it, and don't be ashamed of it, either.

6. *Stay businesslike.*

You may like writing but you are bound to like other things as well. Fine, but if you want to be *prolific,* then just train yourself to like other things less well.

It is very easy to postpone writing because you want to watch football, or because you want to go skiing, or because you want to take a long walk and get close to nature, or because it's time for sex. I don't say that you shouldn't take time out, because such activities and, indeed, any activity, may serve as a source for ideas. There is, however, a fairly sharp limit to how much of this you can do. The law of diminishing returns sets in, and the ideas you get do not make an adequate return for the time you lose.

You have to train yourself to be reluctant to take time away from your writing and to be anxious to return to it. This is not as difficult as you think. The more you learn to make writing easy, the more you develop methods for coming up with ideas, and the more successful the results are in terms of self-satisfaction, sales, and earnings, the more unwilling you will be to stop writing and the more delighted you will be to return to it.

What's more, despite still another stereotype, you can't indulge in alcohol and drugs and still be a *prolific* writer. You can only write decently when your mind is unfuzzed, and if you leave yourself little unfuzzy time, you will end up with little written.

In short, if you can't make yourself treat writing as an almost priestlike calling, and can't lure yourself away from the fleshpots of the world, then you may be a writer—even a good writer—but you will never be a prolific one.

7. *Make efficient use of time.*

I might summarize by saying that everything depends on the attitude toward time.

You can replace money if you lose a wallet, buy a new typewriter or word-processor if your house is burglarized, marry again if a divorce overtakes you, but the hour that has vanished unnecessarily will never return.

There are a variety of ways of using time to the fullest, and every writer chooses his own. Some become completely asocial, tearing the phone out of the wall and never answering mail. Some establish a family member as the dragon to stand between themselves and the world. Some turn off their senses and, while living as part of the world, learn to ignore all that goes around them.

My own system is to do everything myself. I have no assistants, no secretaries, no typists, no researchers, no agents, no business managers. My theory is that all such people waste your time. In the hours it takes to explain what you have done, to check what they then do, and to point out where they did it wrong, I can do at least three times as much myself.

But now that you have my rules for being a prolific writer, do you still want to become one? Apparently, you have to be a single-minded, driven, nonstop person who lets all the good things of the world go by, and who's a pain in the neck to his wife and children, if he's foolish enough (like me) to have them.

So I suppose my final piece of advice ought to be: Don't be prolific. Just concentrate on being a good writer and leave prolificity to those poor souls who can't help it—like me.

(NOTE: Isaac Asimov at this moment has about 15 more books in press, and is typing furiously. His current novel, *Foundation and Earth,* made all the best-seller lists, he was delighted to discover.)

38

FACING UP TO IT

I was asked by an editor to write an essay on some medical emergency in my family that had been bravely met. The only thing I could think of was Janet's mastectomy, but I said that I could not write of it until I was married to her so that I could insure the "happy ending."

Once circumstances permitted marriage, we were married at once, and the editor's request was renewed.

I agreed on the condition that Janet did not object. This was by no means a sure thing. Janet is (unlike myself) an intensely private person, and a mastectomy is not a thing a woman likes to trumpet before the world.

I told her that many women had to undergo it and that our story might hearten others who faced the possibility, and Janet, who is after all a physician, was visibly moved at that.

I then said, "Besides, wouldn't there be something novel about the description of an emergency like that without any mention of prayer and reliance on the supernatural—because there was none?"

She at once agreed.

Incidentally, radical mastectomies are far less likely to be performed now. "Lumpectomies" have become the thing, which preserves much of the breast and may be every bit as life-preserving as cutting out everything in sight. However, there is nothing to be done about that for us. And we are, in any case, after thirteen years of marriage, still very happy with each other.

A lump anywhere on the body, in these cancer-conscious times, makes for nervousness. A lump on the breast makes for sheer panic.

And that's what my fiancée found she had in the spring of 1972—a lump on the left breast.

She is a doctor and knew a good deal about what it could be, and what consequences might follow, and what could very well have to be done—and none of it made her feel any better. There was, however, nothing to be done but to attempt to follow a normal life, professionally and personally, while the matter was investigated.

The various doctors were hopeful. Not every lump is malignant, and there seemed no clear signs of malignancy in the mammogram. Just the same, the lump, whatever it was, would have to come out.

On July 23, 1972, she entered the hospital and I spent considerable time with her there, trying to share with her the confidence I felt. It was nothing, I assured her; she'd be out in no time.

At 3:00 P.M. on July 25, she was taken down to the operating room, quite cheerful, and I sat down to wait. I expected her to be back by 5:00 P.M., and by 6:00 I began to grow anxious. I could get no information out of anyone, other than the fact that she was still in the operating room.

Anxiety grew, hope dwindled, and by 8:00 P.M. I could find no way of consoling myself any further. I knew what was happening—and it had indeed happened.

The lump had shown signs of a small area of malignancy, and a radical mastectomy had been performed. By the time I finally saw her at 1 A.M. she was woozily awake and knew what had happened.

"I'm sorry," she said. She was *apologizing* to me.

"It's all right," I said. "We couldn't let it stay there, could we?"

She closed her eyes and drifted off again, and the doctor, pointing firmly to the door, sent me home.

The next day, she was pretty much under medication and not quite able to grasp the full implication of what had happened, but when I visited her on the twenty-seventh, it had finally hit home and she was in tears.

It is important to understand her position. She was forty-five years old and had never been married. The chief reason for this had been her intense pursuit of a medical career, but she had no great confidence in herself as a classic beauty.

I represented the closest and most nearly lasting relationship she had yet formed with a man, but we were in no position, at the moment, to marry; nor would we be in such a position for a year at least, perhaps longer. In depressed moments, even before the lump had appeared, she felt that I would tire of her and move off in pursuit of younger and more beautiful women. Nor were my assurances to the contrary particularly helpful, since we all know what deceivers men are.

How could she feel, then, lying in bed, scarred and feeling deformed? Her breasts were small in any case, something hard for a woman to accept in our breast-conscious society, and now one of them was gone. Not only did she feel that I could not possibly want to stay with her any longer, considering that no legal bond held me, but she felt that it was necessary for her to make it easier for me to leave by *telling* me to go.

For a while, I could only sit there, hold her hand, and mutter the usual litany of the man who, under such circumstances, feels his love and loyalty unshaken. I explained that I still loved her, that the missing breast didn't matter, that what I valued in her was beyond the reach of the surgeon's knife. It didn't help, of course. She was in no mood to listen to anxious reassurance.

So I stood up, rather desperately, pointed my finger at her and said, "Listen, what's all the fuss about? If you were a showgirl, I could see where taking off the left breast would be tragic. You would be all unbalanced and you would fall over to the right side. In your case, it scarcely matters. In a year, I'll be looking at you and squinting my eyes and saying, 'Which breast did the surgeon remove?'"

And to my relief she burst into laughter. Indeed, when her surgeon arrived to examine her, she told him what I had said and he burst into laughter, too.

It seems, looking back on it, a rather cruel remark to have made, but it seemed necessary at the time. I simply could not allow her to build the mastectomy into a tragedy, and the only alternative I could think of was to take it lightly.

During the remainder of her stay in the hospital, she was in quite good spirits. She was on a floor where the women were, by and large, undergoing hysterectomies, something she herself had been spared. She could walk freely, which they could not do. So she visited them and boosted *their* spirits.

But she was a doctor, after all, and the hospital atmosphere was rather homelike to her. I knew that once the hospital routine was over and she came home to the one-breasted life that would then stretch before her indefinitely, there would be a considerable let-down.

For one thing, she would have to face the necessity of letting me see her as she now was—sooner or later. She would be convinced that the sight of the scar and the "deformity" would disgust and repel me and make any physical affection between us impossible.

She would have to think so, for she knew how queasily I reacted to unpleasant sights.

I knew that, too, and I also knew that I must not let her think I considered her an unpleasant sight. So I called the surgeon and asked him to describe to me exactly what she would look like as a result of the operation. I asked for the location of the scar and the length and direction and redness and whether the needle-marks would show. He warned me that the pectoral muscle had been removed and that the ribs would be in plain view under the skin.

Eventually the bandages were removed, and for a period of time I let her carefully conceal herself at those times when she might normally not have been concealed. I had concentrated hard on imagining what it might look like and finally I seized an occasion to remove the wad of clothing she was carefully clutching to her breast.

I looked at the scar without wincing, and with what I hoped was an absolutely expressionless face, and said, "All right."

I watched her exercising; I went with her when she shopped for prostheses; and I remembered, periodically, to stress the fact that there was no point in making a tragedy out of what could not be helped.

Once a couple who knew of the operation was visiting, and they were busily engaged in making conversation very wide of the mark, carefully saying nothing that could possibly give rise to the thought of breasts.

"Have you ever been at a swinging singles bar?" said one of them to her apropos some comment on contemporary mores.

"*Been* at a swinging single?" I put in at once. "She *has* a swinging single."

For a moment I thought I would be assaulted by the infuriated couple, but my girl stopped them by saying, "Oh, that's just his flattery. It's not big enough to swing."

She was learning to take it lightly, too.

Of course, we would rather it hadn't happened, but the choice was not between losing or keeping a breast; the choice was between losing a breast and losing a life. I would far, far, *far* rather have her without a breast than not have her at all.

And by a mixture of constant reassurance and lighthearted reference, I convinced her of that. I don't know that this same strategy would work with everyone or even with very many, for every person is an individual, but it worked with her.

She no longer makes any effort to hide the "deformity" from me, and, in fact, she feels sufficiently at ease with the situation to let me write this article.

And she didn't lose me. We were married on November 30, 1973.

Additional Note, by Janet Asimov

There's something I must add to what Isaac has said. Beyond the fear of deformity and of the loss of love is something more universal: the fear of death. The mastectomy patient has had her brush with cancer, and there can never be complete confidence that the victory, won at such great cost, is final, or can even long endure.

Death is the lot of us all, of course, and we are all, ultimately, its victims. Most human beings manage to avoid awareness of this, but anyone whose tissues have, anywhere, been touched by cancer, loses some of the ability to do this. Death has left his calling card.

Mastectomy patients therefore live with more conscious uncertainty than do most of their friends, and worry over all sorts of physical symptoms that once they would have taken in stride. Especially during the first few years, it is so difficult to avoid hypochondria—and yet that must be avoided as far as possible if life is not to degenerate into a morass of imagined ills.

The patient should try, over and over, to accept the basic uncertainty of life. She must avoid the kind of isolation that would lead to a too-ingrained self-pity. She must avoid hiding the fact of the mastectomy as though it were a disgrace or crime, for that would lead to a too-ingrained self-disgust. It helps to talk to other mastectomy patients—through the "Reach to Recovery" movement, for instance.

Yet if near-collision with death makes us bitterly aware of the sadness and harshness of living with a future out of control, it may also teach us to live more in the present. If we commit ourselves to being fully alive, if we feel and savor each passing moment as intensely as possible, it will turn out that there will be so many moments neither sad nor harsh, that in balance life will not have been damaged by the experience after all, but improved.

---- 39 ----

TRIPLE BYPASS

My odyssey through the realm of heart disease began on May 9, 1977, when, while engaged in carrying out my morning errands, I stopped midway with a dull pain in my chest and a feeling of being unable to breathe. My puzzlement lasted only a few seconds. I knew what it was: angina pectoris.

My father had developed angina at the age of forty-two and had then lived with it for thirty years. I was superstitious enough to think I would get it at forty-two also, but I was wrong. I waited fifteen additional years, and when it came I was fifty-seven.

I didn't say anything for a few days but tried to puzzle out what I ought to do. On May 18, the decision was taken out of my hands when I had a mild heart attack. (A mild heart attack is one you survive.) I was on a lecture tour at the time and had a deal of other work, and didn't get to my internist, Paul Esserman, for eight days.

Paul attached the terminals, started the electrocardiogram (EKG) going, and the expression on his face after the first second of operation told me exactly what I didn't want to know. He said I would have to be hospitalized right away, and I said I had to give a commencement address at Johns Hopkins in two days and should be hospitalized only after my return. The contest of wills ended with my being admitted to the hospital's intensive care unit inside of half an hour.

I stayed in the hospital sixteen days, then remained under house arrest for two weeks more, and then (as far as I was concerned) I was well. I occasionally had an anginal sensation, when I walked too quickly, or too far, or too tensely, but if I waited ten seconds (I timed it) the feeling went away, and if I then walked more slowly and less tensely it did not return.

That went on for six years and never once was it necessary for me to take medication. I didn't even bother carrying nitroglycerine tablets with me in case of emergency. I was sure there would be no emergencies.

Then came the day of reckoning, for during all those six years my coronary arteries had been narrowing because of the buildup of atherosclerotic plaques. I became aware of this on August 24, 1983, when Janet and I walked to the theater in Central Park to see *Non Pasquale,* and I could barely make it. During the course of the six years I had walked there and back (roughly a mile each way) with little trouble, but this time I had to stop every few blocks and it took far longer than ten seconds for the pain to stop.

I told myself it was the result of tension and of the fact that I had allowed myself to gain a little weight. All I had to do was to lose some weight and cultivate a free-and-easy attitude and all would be well. I began the process of weight loss and did a lot of smiling, but it did no good. My anginal moments continued to increase in frequency and severity.

Still, I was stubborn. During my September session with Paul I said nothing. By October, however, my optimism could no longer be maintained. On October 17, I broke the news to Paul, laughing lightly as I did so to indicate I didn't think it was serious. Paul did not laugh lightly. He listened, got on the phone, and made a date on my behalf with his favorite cardiologist.

On October 21 I met Peter Pasternack, the cardiologist, for the first time. I told him my story. Peter listened to my heart, ran an EKG, and told me that what I needed was a stress test. The best laboratory for the purpose was only a few blocks from my apartment.

On October 26 I walked the fifth of a mile between my apartment and the laboratory with great difficulty and had my stress test. It involved the injection of a radioactive isotope into my bloodstream. That would enable pictures to be taken of my heart in motion, while I walked on a treadmill. Since I don't view the insertion of needles into my veins with equanimity, I thought it best to warn the doctor in charge that if he let anything happen to me he would be torn to pieces by my maddened fans. He informed me that *he* was a fan of mine and if he let anything happen to me, he would kill himself. (I wasn't sure that would help me any, but I let myself be consoled.)

I flunked the stress test. It turned out that my blood supply went down with exertion, rather than up as it should have. This meant that under exertion my heart received less blood through the coronary arteries than it did at rest, although it needed *more*. The result was anginal pain. The heart itself, fortunately, was in good shape. It had not been significantly damaged by the heart attack of 1977, and only the coronaries were shot. That was small comfort. The heart, however intact, won't work without an adequate supply of blood through the coronaries.

Peter went over the stress results on October 29, and said we would need exact figures. That meant an angiogram. In an angiogram, a thin catheter is inserted into the femoral artery at the point where the thigh meets the torso. It is worked upward to the heart. A dye is injected through the catheter into the heart and photographs can then be taken which show the exact places where the coronaries are blocked and the exact degree of blockage.

I lay on the operating table on November 15, in an agony of apprehension, and said, "Let me know when you're going to insert the catheter so I can steel myself."

The doctor in charge said, "Too late. It's half-way to your heart. You can see it on the screen."

I didn't look. It wasn't my idea of an exciting adventure film.

On November 25 I had another session with Peter Pasternack and he gave me the results. Of the three coronary arteries, the largest was 85 percent blocked, the middle one was 70 percent blocked, and the smallest was 100 percent blocked.

"You've got enough blood supply to carry you through under sedentary conditions, so if you wish you can carry on without surgery, making use of medication to relax the blood vessels. You may live out your full life span but you couldn't do much but sit around or walk slowly. You'd be a cardiac cripple."

"And the alternative?" I asked.

"The blockages are near the aorta in each case, and the vessel beyond is in good shape. Each blockage can be bypassed."

"What is the death rate in such a bypass operation?"

"About one in a hundred," said Peter, "but that includes everyone—old people, people with bad hearts, emergency operations, and so on. In your case, it should be considerably less than one in a hundred."

"And if I decide to do nothing, what are the chances that I'll be dead of a heart attack within a year?"

"About one in six, I should judge," said Peter.

Without hesitation I said, "I'll take a triple bypass."

"I think you're wise," said Peter, "and I'll arrange to have you operated on by Stephen Colvin. I had him operate on my mother last year."

I'm not a mystery writer for nothing, however. I saw the loophole in the argument. I said sternly, "Do you love your mother, Peter?"

"Very much," he replied.

"Then I'll take Colvin."

On November 29 I met Stephen Colvin, the surgeon. He was a thin, lively, very intense young man, who suited me completely. He was just the right age, old enough to have done innumerable bypasses, young enough to have a rock-steady hand. What's more, he clearly lived, breathed, and ate bypasses and was interested in nothing else. I judged him to consist of little more than an arm to hold a scalpel and eyes to guide it, and that was exactly what I wanted.

He said he could fit me into his schedule in two weeks, unless I wanted to wait till after the Christmas/New Year holiday season.

Actually, I desperately wanted to wait because on January 6, 1984, there would be the annual banquet of the Baker Street Irregulars (a group of Sherlock Holmes enthusiasts) and I was customarily the last item on the entertainment program, when I would sing a comic song of my own devising. However, I didn't think it was safe to wait and the operation was scheduled for December 14.

As it was, there was a significant chance that I might not make it. On December 6 I forgot myself and ran for a taxi. Adrenalin kept me going while I ran, but once inside the taxi, the worst anginal attack I had ever had over-whelmed me. I was convinced that it would end in a killer heart attack and that the driver, rather than be stuck with a dead body, would drive to the river and dump me there, and that my wife, daughter, and fans would never now what had happened to me. I was about to write a little note, "I am Isaac Asimov," and put it in my jacket pocket, when the angina eased up. After that I took to carrying nitroglycerine tablets and taking them when necessary.

At 3:00 A.M., December 2, unable to sleep, it occurred to me that I had not written my song for the Baker Street Irregulars banquet. Then and there, I made one up to the tune of "Danny Boy." Here it is:

Oh, Sherlock Holmes, the Baker Street Irregulars
Are meeting here to honor you today,
For in their hearts you glitter like a thousand stars
And like those stars you'll never pass away.
This year that's new must tick away its months and die,
For Father Time moves on remorselessly.
But even he can't tarnish as he passes by,
Oh, Sherlock Holmes; oh, Holmes, your immortality.

Oh, Sherlock Holmes, the world is filled with evil still,
And Moriarty rages everywhere.
The terror waits to strike and, by the billions, kill.
The mushroom cloud is more than we can bear.
And yet there's hope in what you've come to symbolize,
In that great principle you've made us see.
And we may live if only we can improvise,
Oh, Sherlock Holmes; oh, Holmes; your rationality.

A few days later, I sang it onto a cassette. It was my idea that, since I would surely be unable to attend, my dear wife, Janet, would bring it to the Hotel Regency and have them play the cassette at the banquet there. The banquet is stag, but I felt they would stretch a point under the circumstances and let her enter.

On December 13 I entered University Hospital, the best possible rendezvous for bypass victims (with the possible exception of Mass General in Boston). The anesthesiologist came to see me and I asked him a question that had been bothering me for weeks.

"Listen," I said, "the only way you can insert a bypass is to make a hole in the aorta. The instant you do that a torrent of blood emerges and I die. Isn't that so?"

"Didn't anyone tell you?" he said. "First we stop the heart."

I turned green. "But if you stop the heart, I'm dead in five minutes."

"Not at all. We put you on a heart-lung machine and it will keep you going for as long as necessary."

"What if something goes wrong with it?"

"Nothing will go wrong with it. Even if there's a citywide blackout, we'd keep on going on our emergency generators."

"What if the heart won't start again when the operation is over?"

"Not a chance. It *wants* to start. The difficulty is keeping it stopped."

I had to be satisfied with that.

That night, a villain with soapsuds and a razor shaved my body from my chin down to my toes, removing (to my horror) even my pubic hair. My only comfort was that he left my sideburns intact.

At 1:30 P.M. on December 14 I took three pills that were intended to tranquilize and sedate me. These things work perfectly on me. Within twenty minutes, I didn't have a care in the world. My dear wife, Janet, went with me as I was wheelchaired to the elevator en route to the operating room, and from that point on, till 10:00 A.M. the next morning, I remember nothing.

I remained responsive, however, till they applied the anesthetic, and I was told later that I wouldn't let them give me the anesthetic till I had sung a song for them.

"What song?" I asked.

"Something about Sherlock Holmes," I was told. Obviously, the banquet was on my mind.

The operation took six hours and I was told it went "better than average." I said as much to Steve Colvin when he came to see me in the recovery room.

"Better than average!" he said indignantly. "It was *perfect*."

Actually, they made use of a mammary artery to bypass the largest of the coronaries. This is better than a vein, for arteries are wider and stronger, but the mammary can't always be used in patients my age. Fortunately, my mammary proved to be in excellent shape. A vein from my left leg was used for the other two bypasses. I ended up with a ten-inch scar down the middle of my breastbone, and a twenty-inch scar down my left leg from midthigh to ankle.

One other event of interest took place in the course of the operation. I don't remember it, but Paul Esserman swears to it.

Once I found out I was going to be on a heart-lung machine, I worried about them supplying my brain with enough oxygen. Even a small and temporary shortage might take the keen edge off my mentality, and while in ordinary people that mightn't show, in my case I was convinced it would. I hesitated saying this to the anesthesiologist, but I explained the situation to Paul.

"Don't worry, Isaac," he said. "I'll keep them aware of your needs, and after the operation I'll test you."

After the operation, I would occasionally come out of my coma briefly and then fall back into it. As I said, I remember nothing of those moments. Paul was there, however, on the occasion of one of these temporary recoveries. My eyes fluttered open, I saw him, and muttered, "Hello, Paul."

At once he leaned forward and said, "Isaac, make up a limerick."

Whereupon I said,

> There was an old doctor named Paul,
> With a penis exceedingly small. . .

And he said, austerely, "That's enough, Isaac. You pass."

In the recovery room, at 10:00 A.M. the next morning, when I really came to, I heard the *New York Times* being hawked. I had no money to buy one but a kindhearted nurse bought one for me. I proceeded to read it with great interest, for I hadn't been that sure there would be a December 15 for me. A passing doctor said, "What are you doing?"

"Reading the *New York Times*," I said. (What did it look like I was doing?)

He said indignantly, "I never heard of anyone reading a newspaper in the recovery room."

"I'm the first," I said, calmly, and kept on reading.

I stayed in the hospital until December 31. The first week I was in the hospital proper, in a private room, with private nurses round-the-clock. (Poverty is no disgrace, but it's very convenient to have money.) The second week I was in the Co-op Care Unit, which is very much like a hotel room, and there my dear wife and my beautiful daughter took turns staying with me.

I attended teaching sessions in which we bypass patients were told that recovery would take two to three months, that there would be ups and downs, and that we would go through periods of depression.

I wasn't worried. Once I got home, I knew, I would make straight for my typewriter and word-processor and that was all I cared about.

That, and one other thing. I laid siege to Peter Pasternack, to let me attend the Baker Street Irregulars banquet.

"Well," he said, "if the temperature is above freezing, if it isn't raining, and if you're feeling well, you can go."

It didn't sound as though there would be much chance. December had been the coldest December on record, and I imagined that the deep freeze would continue.

And yet January 6 arrived on a mild wind. The evening temperature was in the forties, there was no precipitation, and I felt pretty well (and said I felt *great*).

Janet and I took the cassette in case it turned out I couldn't sing, and we also took a taxi. I arrived just in time for the intermission, which meant I had a chance to meet all my B.S.I. cronies, and listen to them tell me how well I looked.

Julian Wolff, the Ayatollah Khomeini of the Baker Street Irregulars, put me on as soon as the intermission was over. I made a little speech describing my operation, then sang my song in a rather cracked voice, and got a tremendous ovation.

I went home in seventh heaven, having beaten the odds and the operation, and having done my bit, as planned. After that, recovery was a piece of cake.

Afterword: There is a certain insensitivity about me. Since I don't have any particular feeling of privacy myself, it never occurs to me that anyone else does.

My first stirring of horror after the publication of the previous essay came when Paul Esserman told me that many of his patients were sending him copies with the two lines of the limerick underlined.

"Oh, my goodness, Paul," I said, "do you mind?"

"Not at all," he said, grinning (for he's the sweetest fellow in the world), "but I intend to sue you for patient malpractice." But he never did.

On my next visit to Peter Pasternack, I found out he had been sent copies, too, but he said, "Thank you. You made my mother very happy."

"I did?"

"Yes, you arranged to have it said right there in print for all her friends to see that her son loved her."

"How about Steve Colvin?" I asked. "Do you suppose he's annoyed at being described as nothing more than an arm to hold a scalpel and eyes to guide it."

"No," said Peter. "He was flattered. That's the way he sees himself."

So all went well, and now three years after the bypass I continue to feel great.

One more thing. I don't like to make a big tragic deal of my misfortunes, but it did give me a shock to hear someone say, "I read your essay on your triple bypass. It had me laughing all the way."

PART III
FUTURE

40

THE ELEVATOR EFFECT

Prophecy is a mug's game, because there's no way it can allow for the unexpected, except by guessing.

The result is that prophets almost never see what, later on, turns out to have been obvious.

Here is an example of what I mean—

In the course of the half-century period between 1919 and 1969, innumerable science fiction stories were written about flights to the Moon, some of them pretty knowledgeable about the requirements of rocket flight and the physical conditions on the Moon.

Up to 1948 a considerable number of stories were written about television; after 1948 some were written about communications satellites.

Not one story up to 1969, however, ever put these three factors together and clearly predicted that when the time came for the first landing on the Moon, hundreds of millions of people on Earth would be watching it on television.

As far as I know, no turn-of-the-century imaginative thinker who pictured a future world in which automobiles were common ever thought of such things as the parking problem or air pollution. No one who speculated on the taming of atomic power ever dreamed of the problem of radioactive ash disposal. Of all those who speculated during World War II on what the postwar world would be like, I can't recall one who predicted that the African colonies would start becoming independent within fifteen years of victory.

I call this the "elevator effect" for the following reason.

Suppose this were 1850 and I were trying to picture the city of New York a century later. In order to help me, some kindly magician had given me a quick glance of a photograph of twentieth century Manhattan that he had brought back in time for me, so I therefore knew for a fact that the island would contain many buildings in excess of ten stories in height and that at least one would be one hundred stories high.

Given that, it would be my task to picture the details of a city in which this was so, using my 1850 experience for the purpose.

In the first place, it would be clear to me that climbing more than six or seven stories would be an arduous task and that therefore people who had made it to the upper reaches of a skyscraper would be reluctant ever to leave it. Each skyscraper would therefore have to be as self-contained an economic unit as possible.

It would have to contain food stores, tailors, barbers, taverns, gymnasiums, and all the other appurtenances of civilized living—all duplicated periodically as one went up the skyscraper. Supplies would have to be hauled up by pullies operated by steam engines and those supplies would be taken in at special doors placed in the side of the building at various levels. There would be no time of day or night when supplies of one sort or another were not being hauled up the outside of a number of skyscrapers.

There would be bridges from building to building at various levels, too, so that people could travel in a limited way without too much moving up and down.

Suppose people occasionally had to leave the building for social or business reasons. Those who lived in the bottom five stories would have little trouble and would be expected to pay premium rents in consequence. What of the proletariat in the upper stories, however?

There would be a spiral slide down which people could descend. It would be undignified and might give rise to vertigo and nausea among the unacclimated, but to the tower-dwellers it would be an accustomed thing.

Getting back up to the higher reaches of the building, however, would be very likely a half-day's journey. The wise person would allow for that and would stop every five stories at the lounge provided for the purpose, sit down, have a small drink, glance at the newspaper. Eventually he would make his way back home.

One could go on and on, building a rational picture in greater and greater detail, describing how the skyscrapers are built, of what materials they would be constructed, and so on.

The question is, though, would I think of the elevator? If I did not, the entire prediction would be wrong, completely wrong, ludicrously wrong.

And I think most people would not think of the elevator.

Well, then, what will New York be like forty years from now, in 2020 or thereabouts?

That depends, doesn't it, on what decisions humanity as a whole will make? If humanity decides on a thermonuclear war, New York City will, in all likelihood, be a weakly radioactive desolation forty years from now. If humanity decides on wasting its way to the end of its oil supplies without producing sufficient supplies of an alternate energy source, then New York might stand more or less in disarray, inhabited by hostile street gangs scrounging for what they can out of what is left of the city.

These are things that are to be expected if humanity behaves in an insane (or even merely in a foolish) way.

I want to look for the unexpected, however, and try to take into account the elevator effect.

One unexpected eventuality would be to have human beings behave sanely and rationally, so that there would be no thermonuclear war and no foolish dispersal of our energy reserves. We can picture instead a reasonable cooperation between nations and an international push toward energy conservation and toward the development of a more advanced energy technology.

This would not be the result of any sudden growth of love and brotherhood (we might seek for the unexpected, but it would be no use hoping for miracles). It would be the result, instead, of a common fear of death and destruction, and of a common groping for survival. Cooperation for the sake of survival is unexpected enough and could be a completely adequate aspect of the elevator effect.

Under these conditions, we can expect forty years of continuing technological advances, and among these advances would be such things as:

1. A continuing advance of computers in the direction of increased versatility, miniaturization, and availability, so that the home computer would be as common in 2020 as the home television set is now.

2. A continuing advance of communications satellites, and the use of laser beams for communication in place of electric currents and radio waves. A laser beam of visible light is made up of waves that are millions of times shorter than those of radio waves and have room for millions of times as many channels for voice and picture. Every person could have a different TV channel assigned to him.

3. An international effort in the exploitation of space, of building solar power stations in orbit around the Earth, as well as observatories, laboratories, factories, and space settlements. Forty years from now this push would still be in its early stages, but it would be visibly progressing, and it would encourage people to think of themselves as Earthmen rather than as this nationality or that.

4. Nuclear fusion would offer a controlled and practical source of unlimited energy, and there would already be important fusion power stations in existence.

In addition to these technological advances there would be a social change of great importance—a dropping of the worldwide birthrate.

This would come about for a number of logical reasons. First, there has been so much talk about the dangers of an endlessly increasing population that the governments of the world have for the most part become aware of this and are doing their best to encourage such a drop.

Then, too, before things get better, they will become worse. In 2020, the population of the Earth will be perhaps seven billion, while energy will be in short supply. Oil will be produced in quantities far behind need, and the substitution of shale oil and coal will not make up the gap. Though nuclear fusion power stations will exist, they will not yet be producing enough energy. And the real deployment of solar power will have to wait for the building of a number of collecting stations in space.

Times will therefore be hard on Earth. However, there will be the prospect of a period of declining population and of rising energy, so couples would be motivated to postpone having children.

Such a drop in birthrate took place in the United States during the Great Depression, but it was then deprecated and people spoke of "race suicide." The forthcoming drop in birthrate will be applauded and encouraged.

Put all this together and what is the conclusion that can be reached? Skipping over all the obvious and getting to the bottom line at once, it is that cities are going to be obsolete.

For ten thousand years, the world has been getting urbanized at a faster and faster pace; at a breakneck pace, indeed, since World War II. We're coming to the end of that stage of the game, however.

Since this is a consequence of technological advance, the trend will be most noticeable in those areas of the world that have been most technologically advanced for the longest time—the northeastern United States and the Midwest—and New York City will be in the lead.

There is a beginning even now. The older American cities are decaying and losing population. We can see some of the more obvious reasons for it. Since World War II, poor people have been flooding into the cities seeking the amenities of the city: jobs, welfare, excitement. At the same time, the middle class has been moving out of the cities and into the suburbs. Rising welfare costs and declining tax bases are reasons enough for urban decay, but they are only the first symptoms of urban obsolescence.

The reason that cities exist at all is to bring together in one spot a great many people so that together they can do what in isolation they could not. The original cities consisted of people gathering together behind a wall, for mutual defense against marauders.

Eventually, it turned out that cities made specialization possible. Artisans and artists could prosper in a crowded and stimulating environment. Wealth and prestige accumulated with the crowds. The city became a center of religion, business, literature, and art—all because people could reach people and interact.

Through most of history, however, people could only reach each other and interact when they were actually, or at least potentially, in physical contact. They had to be in reach of arm and voice, and so they gathered into a clot. As technology advanced, larger and larger conglomerations could be successfully fed and served—so the cities grew larger.

By the mid-nineteenth century, however, long-distance communication arrived in the form of the telegraph, and this was followed by the telephone, radio, and television. Easy long-distance transportation arrived in the form of the railroad, and this was followed by the automobile and the airplane. People could reach each other quickly even when at great distance—by voice and image in seconds, by actual physical transport in hours.

The necessary contact and stimulation could, by the twentieth century, stretch across continents and oceans at will. Why, then, should people continue to accumulate into physical masses and the cities continue to grow?

In the first place, technology advances irregularly on Earth, and there are regions where long-distance communication and transportation have not yet reached the mass of population, so cities still have positive and irreplaceable advantages.

Second, old habit and old tradition is strong. We all know that time-honored institutions live on long after their real reason for existence is gone. We might mention the monarchy in Great Britain and the electoral college in the United States as examples (though, of course, plausible reasons for their continued existence are easily enough invented by ingenious people who prefer the old to the useful).

Finally, the population of the Earth has been rising steadily and even more rapidly in the last century than in previous centuries. Therefore, countering those factors that are pushing the city into obsolescence are the sheer crowds of people from the countryside pushing into the city for its advantages, or supposed advantages. Thus, the cities continue to grow even now, over most of the world, in increasingly distorted and miserable fashion.

Yet despite all these factors that tend to allow cities to survive and even grow, we see the beginnings of the end in the United States. The trend is to the suburbs, and why not? There is no longer any disadvantage to the suburbs. You can be out there and yet still be "with it." Forty-five minutes from Broadway makes no difference any more. For that matter, forty-five minutes from Broadway is no longer New Rochelle, as it was in George M. Cohan's day, it is Boston.

Is this a bad thing? No, it is a natural thing, and it is a trend that will continue.

Just as the elevator made it possible for cities to expand upward, so increasing ease of communications is making it possible for cities to evaporate outward.

If communications satellites make it possible for each person to reach any other person by sight and sound and, perhaps, by three-dimensional holography; if factories and offices are so utterly automated and computerized that they can be as easily guided and controlled from a distance as they can by someone on the spot; then anyone can live anywhere on Earth without undue penalty.

Wherever you live, you will be able to see anyone, at least by way of three-dimensional images, without moving from your home—either for social or business reasons. You can have conferences with any reasonable number of individuals, all in the same conference room if you count by images—even though the corporeal bodies are separated by continents. Any document, article, newspaper, or book can be brought to you, and any cultural event can reach you. The readings of any monitors can be flashed on your screens, and your commands can as easily influence the workings of a distant office or factory as scientists in Houston can control the digging arm of the Viking space vessels on the surface of Mars.

If you must see someone in the flesh, if you are anxious to stare upon an authentic wonder on Earth's surface, if there are cases when images don't satisfy you, you can travel in the old-fashioned way. And you can do so all the more comfortably since most people will *not* be doing so at any given time.

In that case, why live in some special place only because several million other people also live there? Why not live where you want to? Mere distance will impose no business, social, or cultural penalty. With transportation facilities not clogged with people, and with their thorough computerization and auto-mation, it should not be difficult to get material necessities to you wherever you live.

Will all this come to pass in forty years?

I think not to completion, but the trend could well be unmistakable by then. With distance no longer a factor, and with population on the point of a long-term decline, city populations will be thinning out. (To be sure, humanity will be beginning to move into space settlements, so total numbers will someday be moving up again—but probably not on Earth itself.)

New York City in 2020 will be opening up—open spaces, that is. The slums will be undergoing an uprooting process and will disappear, giving place to open land, to parks, gardens, small bits of farmland.

This is nothing to cry about. If what we call cities begin to wither and disappear, it will be because (if we stay sane) the whole planet will be in the process of becoming a single city—well-scattered over the face of the world, scattered among parks, and farms, and wilderness, and with the space settle-ments as its suburbs.

And as the decades and centuries continue to pass, the space settlements will spread out more widely, to the asteroid belt and beyond—to the farther planets and beyond—until finally, perhaps, all of Earth may be obsolete, and humanity will be moving out to occupy the Universe, and to take its place with such other advanced civilizations as may have stepped out of the cradle of planethood.

41

2 0 8 4

Here we are in 1984, an ill-omened year since 1948, when George Orwell, in the book that names our year, pictured a world in the unremovable grip of a vicious tyranny.

Fortunately, Orwell's world has not turned out to be the actual world of 1984 that we are living in. There are tyrannies on Earth in nations both large and small, it is true, and no nation exists without its injustices. However, there is freedom as well, and even the worst tyrannies of the real 1984 find they must take the human yearning for liberty into account.

Let us, therefore, with considerable relief, look beyond the present into the possible world of 2084.

There is no use denying that our feeling of relief cannot be total. There may well be no world worth describing in 2084. Humanity does have the power to destroy itself. Nuclear war might break out tomorrow. Overpopulation might reduce us to starvation in decades. Pollution and the mishandling of Earth's resources might make our planet increasingly unfit for human life even if we avoid explosions either of bombs or population.

But on the other hand these dangers are not creeping up on us unannounced. There are strong movements in favor of peace, environmental protection, and population control. And we have powerful technological tools to deal with various dangers, tools that even a generation ago we did not have.

Let us assume, then, that we will avoid catastrophe. In that case, what will the world be like a century in the future?

That is not a question easy to answer. Could anyone in 1884 have made a reasonable guess as to what the world of 1984 would be like? Could anyone have predicted nuclear bombs, giant jet planes, communications satellites, computers, and close-up pictures of Saturn's rings?

Since the rate of technological advance has been accumulating all through history, the chances are that the changes of the next century will far outstrip in scope and unpredictability the changes of the past century. It seems certain, then, that any attempt I make to picture what 2084 will be like will probably sound enormously funny to those who will eventually be living in that year, but I will take the chance, anyway.

It is clear that the dominant technological factor of the society of today is the rapidly advancing computer. At the rate at which computers are improving and taking over, it seems very likely that by 2084 they will be doing *all* work that is routine and repetitious. The world of 2084 will therefore seem to be "running itself."

This has happened before in history. A person from 1784, viewing the world of 1984, observing the machinery that does all the pushing and pulling and digging and lifting at the shift of a lever or the touch of a button, noting how machines move us faster than horses can, and produce writing more clearly than the hand can, how they cool and heat houses, freeze and cook food, and do many other things with scarcely any supervision, would think that we of 1984 live in a world that is running itself.

This shift of work from muscle to machine will now continue with a shift of work from brain to intelligent machine.

What will be left, then, for human beings to do? Only everything—everything human, that is; everything that involves insight, intuition, fancy, imagination, creativity.

To begin with, there will always be the task of designing, constructing, and maintaining the computers and their programs. In addition, there will always be the task of taking care of the human aspects of society: politics, economics, law, and medicine. There will be scientific research, and there will be art, music, and literature.

There will, in fact, be a vast number of creative jobs that we cannot name, or even imagine, today. All through history, the advance of technology has managed to create more jobs than it has destroyed, but it would pass the bounds of the possible to imagine beforehand what those jobs might be, since they would depend upon the particular advances that technology will make.

A person living in 1784, when ninety-five percent of the world's population was involved in agriculture in a fairly direct manner, if told that just two centuries later only five percent would be so involved in the most advanced countries, would be unable to imagine what the rest of the population could possibly be doing. Could they foresee telephone operators, airplane mechanics, astronauts?

One might question whether it were impossible for the vast bulk of the human population to be involved in creative work. Surely, creativity is reserved for a small elite, while billions of ordinary people can do nothing better than plod away at the physical and mental drudgery from which they will have been ousted by the computers and robots of a heartless technology. Will those billions be left behind as hopeless, helpless misfits, unable to participate in society?

This may be so in the short run, for most of the people of the world have grown old without much of an education and have been forced to devote their lives to a kind of meaningless repetitive labor that has permanently stultified and ruined their minds.

However, the potentialities of human beings who are properly educated and who have been supplied with the appropriate mental stimulation through their lives are not to be judged by the ruins of the past. And there are precedents for this.

There was a time, in medieval Europe, when the ability to read and write seemed confined to but a small elite. Undoubtedly, the scholarly few viewed with contempt the brutish peasantry and aristocratic thugs who could not make sense of those small cryptic markings we call letters; this elite would not have believed that the general population could ever be made literate. And yet with the advent of printing, and with state-supported education, mass literacy became possible.

There are new cultural revolutions in the offing. With computerized libraries; with computer outlets in every home that can tap these libraries and other sources at will; with each person able to find at his or her disposal the gathered knowledge of humanity; with all enjoying the privilege of being able to follow programs designed to guide the curious in any direction; with individuals (young or old) able to find out what they want to know at their own speed, in their own time; learning will become fun for all. The twenty-first century will be the first in which the natural creative potentiality of human beings generally will be tapped and encouraged from the first.

Yes! With machines doing machine work, humans will do human work.

Despite all the human work that will remain in the computerized and automated world of 2084, it will be a world in which people will enjoy far more leisure than they have now, and will be better able to make use of that leisure.

For instance, with the further development of communications satellites, the world will be knit together far more closely than it is today. There will be communication channels in the billions, thanks to modulated laser light, and every person can have his own television outlet. Any person will be able to reach any other person in both sight and sound with minimal difficulty. A person from any place on earth can check on, oversee, and control the workings of machinery in offices or factories for which he is responsible. (Naturally, there will have to be strict coding and elaborate security controls, for it is not difficult to visualize computerized crime and vandalism.) People scattered over the five continents can foregather by three-dimensional holographic television.

The result is that the world can be safely decentralized, for it will not be necessary to clump huge populations into small areas in order that they might all be near their jobs, or near cultural outlets either. Commuting, and business travel in general, will undergo a massive shrinkage.

But this will make it all the more possible to travel for pleasure, to meet distant friends or relatives in person, to tour the world and become familiar with it. The pleasure travellers will not be crowded out by the business travellers.

There will be a variety of new modes of travel. There will be air-foils that ride on jets of compressed air, so elaborately paved roads and bridges will become less essential; and those that do exist may be reserved for those vehicles that carry freight rather than people.

There will be trains that travel through evacuated tunnels on monorails that they do not actually touch but over which they are supported by fields of magnetic repulsion. Without friction or air resistance, they will be able to travel at supersonic speeds.

With improvements in communication it seems inevitable that the world will develop a kind of composite speech ("Planetary") that everyone will understand. The facts of life will force this speech to be based more on English than on any other language, but the admixture of non-English elements will make it foreign even to English-speakers. It will be everyone's second language, so that without destroying the richness and variety of the cultural heritage of humanity, it will encourage understanding and friendship.

In the same way, hotels will inevitably become the cosmopolitan meeting place of people from everywhere. Perhaps as computers develop and begin to use and understand human speech, it will be Planetary that they will speak, and hotels will be pioneers in that direction.

Even at home, the combination of leisure time and computerized education will encourage talent so that "show business" will become far more important than it is now. No human being will be considered complete unless he can write, sing, dance, speak—unless he can *somehow* be able to contribute personally to the pleasure of life for others. In the same way, athletics and games will be more widespread. Hobbies will be more numerous and elaborate. It is quite impossible to predict just what the vast varieties of "fun" will be like in a world that will be so alien from us in so many ways.

One thing we can be sure of. Human beings won't have nothing to do in a world that "runs itself"; their lives will be crowded with activities of a kind that will intermingle work and play so effectively as to make it impossible to decipher what's one and what's the other.

In the course of the next century, moreover, humanity will witness a new burst of pioneering, as the steady expansion of the human range, having faltered to a halt as the Earth grew full in the twentieth century, will reignite into a new and unprecedented great expansion into space.

I suspect that in 2084 Earth will derive most of the energy that will keep the machinery going, and allow people to live their full lives of creative work and leisure, from solar power stations in orbit. These will convert sunlight into microwaves which can be beamed to Earth and made into electricity. There may be nuclear power plants in orbit as well.

Most of human industry will be in orbit, too. Factories will be making use of the unusual properties of space—hard vacuum, energetic radiation, high and low temperatures, zero gravity—to manufacture items that can be made on Earth only with difficulty if at all. Food will be grown in huge hydroponic stations in space. All of these will be so thoroughly computerized and robotized as to require little human interference.

Space will represent an enormous sink into which unavoidable pollution can be discharged. The volume available for that will be millions of times greater than that on Earth's surface, and we will not be fouling our nests either, for the solar wind will sweep it away into the incredibly great vastness beyond the asteroid belt.

There will be elaborate mining stations on the Moon to supply most of the needed material to build all these structures, to say nothing of astronomical observatories and scientific laboratories.

There will be numerous space settlements, each holding tens of thousands of human beings, all living in carefully engineered environments. Each settlement may have its own chosen culture and way of life, so that humanity will be developing greater variety and diversity than ever. The space settlements will mimic Earth's environment almost completely, but they are bound to have pseudo-gravitational pulls that vary from place to place. This will have its advantages, too, for imagine the chance to develop skills at tennis and other sports when gravitational pull is closer to zero.

The task of expanding into space will offer an important and desirable goal to the nations generally; one that can be carried through to begin with, and maintained afterward, only by the most generous international cooperation. In 2084, therefore, we may see a federal global government based (it is to be hoped) on the most successful example of federalism on a less-than-global basis—that of the United States of America.

If we can but conquer our own follies, hates, and fears, and refrain from destroying ourselves in the immediate future, the horizons beyond are golden and endless.

42

SOCIETY IN THE FUTURE

It is easy to predict gadgets of the future. Science fiction writers do it all the time. In the past we imagined television, atomic bombs, rockets to the Moon. Now we could look forward to other triumphs of science and engineering.

But what will *society* be like in the future?

Will there be a future society in the first place? Perhaps not, if we really go all out in nuclear warfare.

Even if we keep the peace, will there be a comfortable society, a civilized society, in the future? Perhaps not, if we go on as we are going.

There are over four billion people on Earth. We are having trouble feeding them and supplying them with the services they need. We are destroying the Earth's living space and resources to keep them alive and as nearly comfortable as they would like to be.

At the present rate of population increase, there will be eight billion people on Earth by 2015, and the chances are there will be no way in which our oil-depleted, soil-depleted, resource-depleted planet can support them; so we are facing catastrophe in the course of the next generation.

That's one way of looking into the future of society—catastrophe and a new barbarism.

But will human beings just sit still and let that happen? Or will they, at the last minute, bestir themselves and take action to prevent destruction? Will they reorganize their way of life to allow civilization and comfort (or the chance of future comfort) to survive?

It will take many hard decisions, but suppose humanity makes those decisions. What will life be like in the future?

To begin with, population will have to be controlled. Human numbers must not outpace the ability of the planet to support us. The control can't be brought about through a rise in the death rate, since that is catastrophe (war, starvation, disease, and anarchy are the great death-rate solutions).

The alternative is to lower the birth rate the world over. If that is carried out, women the world over will lose their ancient function as baby-machines. They will tend to have few children—one or two and, in many cases, none.

What are women going to do instead? Nothing?

Women will have to move into the world and take part in all the roles men have so long monopolized—business, science, religion, government, art.

Why not? Humanity needs all the brains it can get, and by making use of women we double the supply without adding to the population at all.

In fact, it may be that only by allowing women to enter the great world can we successfully reduce the birth rate. They will then be less anxious to have many children as the one way they can achieve personal status.

The society of the future will have to be a women's-liberation society—if we are to survive.

In a low-birth-rate society, the percentage of young people will be very low—lower than it ever has been in human history.

On the other hand, if civilization survives, science and technology will continue to advance. In particular, medicine will advance and the life expectancy will continue to push forward. There will therefore not only be fewer young people, there will be many more old people.

Does that mean that the society of the future will be increasingly one of old-age pensions, Medicare, and social security? Will a smaller and smaller reservoir of younger people be required to support a larger and larger dead weight of old people who are retired, pensioned, and sick?

If so, society will break down even if there is no war and even if population is controlled.

But must old people be retired, pensioned, and sick? With advancing medical science, people will surely remain vigorous and capable into advanced years (we have already been moving in this direction for a century).

To keep them mentally alert and creative, there will have to be a revolution in our philosophy of education.

Throughout history, education has been reserved primarily for the young, and it has been delivered massively, to large numbers at a time.

If we are to survive in the future, education must be a lifelong process and it must be individualized. Any human being, at any time, can be educated in any subject that strikes the fancy. This isn't so impossible if we take into account advancing technology.

If we consider communications satellites with laser connections, we can imagine every human being having his or her own television channel on which an advanced and computerized teaching machine can operate. Such a teaching machine could be hooked into a planetwide computerized library containing the reservoir of human knowledge.

Between medicine and computerized education, human beings will remain both physically vigorous and mentally creative into advanced years. In other words, the society of the future will have to be an age-blind world—if we are to survive.

If we do survive, and if science and technology continue to advance, the work of the world will be increasingly done by automation, by robots, by computers. Human beings will have to fall back on *human* activities—on creativity, on the arts, on show business, on research, on hobbies.

In short, the society of the future will be a leisure-oriented society, but not one in which people just kill time. They may well work harder than they do now, but at what they *want* to—if we are to survive.

All of this can't be done simply by people alone. Humanity doesn't live in a vacuum; it must draw on the environment.

Most fundamentally, it must draw on energy sources. But the most convenient source that humanity has ever known, oil, is now drying up.

So there will have to be new sources. There are many small sources that can be used and all should be, but to run *all* the activities of humanity through all time to come there must be at least one large-scale source that is safe to use and will last for billions of years.

There are, actually, two such sources: nuclear fusion and solar power.

The chances are that both will be developed and both used, but there are reasons to suppose that, in the end, it may be solar power that will win out, and that such power will not be collected from sunlight striking the surface of the Earth, but from sunlight striking a collecting station in orbit about the Earth.

In fact, the space about the Earth has a great many desirable qualities. It consists of an infinite supply of hard vacuum, it is exposed to extremes of temperature and to hard radiation, it can be gravity-free. All these properties can be useful to industrial processes, so it would make sense to have factories in space. (Unavoidable pollution is better discharged in space than on Earth's surface.)

There could be laboratories in space, where dangerous experiments can be conducted without risking human populations. There could be observatories in space, where the Universe can be studied without the blanketing and distorting effects of an atmosphere. There could be settlements in space, where social experiments can be conducted, from where the rest of the Solar System can be explored, and through which there will some day be room for population expansion once more. And all could be built out of materials obtained from our Moon.

In short, the society of the future will be space-oriented—if we are to survive.

The problems that face humanity now, and that must be solved if humanity is to survive, face all nations alike. We live in a global crisis and there must be global solutions.

Old and foolish enmities cannot continue; war-machines can be neither used nor even maintained. It is suicide to fight, to prepare to fight, or even to think of fighting. Like it or not, we must all cooperate if we are to escape the precipice.

Population can be controlled; education can be reorganized; space can be penetrated and exploited; energy sources can be used—but only through a worldwide cooperative effort.

In short, the society of the future will be free of war and of racism, and will in fact see the establishment of some form of world government—if we are to survive.

Of course, we needn't suppose that our civilization *must* survive. If human beings would rather have all the children they want, and if they would rather maintain large and elaborate war-machines than develop a new education and devise ways of penetrating space, then they can.

But in that case, civilization *won't* survive. And over the next generation or so, billions will die.

Afterword: I am frequently asked to write essays describing my view of the future. I try to reach a variety of audiences and to stress one aspect or another, but I obviously have a particular view of the future and so there is a tendency to overlap.

For instance, I say some of the same things in the previous essay that I had said in the one before, "2084" (#41), and that I am about to say in the one after, "Feminism for Survival" (#43).

For this, I can only ask your indulgence. Although I repeat myself of necessity, I do so in different ways, and if you find an argument unconvincing in one essay, you may find it more persuasive in another essay when I approach it in a different manner.

43

FEMINISM FOR SURVIVAL

It is easy to argue for women's rights as a matter of justice and equity. Easy, but often useless, for such things as justice and equity are not convincing to those who profit by their absence.

Without in any way denying that there is justice and equity in the concept of women's rights, I prefer to argue for it on the basis of necessity.

It seems clear to me that if we continue to maintain a social system in which half the human race is compelled by reason of irrelevant anatomy to labor at tasks that do not include science, the chances for civilization to endure into the twenty-first century will be sharply reduced.

This should not be difficult to see. We are now facing numerous and weighty problems, and it is obvious that with human numbers growing daily, with the energy supply becoming more precarious daily, with food reserves shrinking daily, with gathering uncertainties producing social unrest and violence that is increasing daily—we are facing a massive crisis which represents life or death to world civilization.

The precise solutions that will help resolve the crisis are not easy to foresee, but we can feel pretty safe in arguing that they will come about, if at all, through advances in science and technology. We *must* have alternative sources of energy, and these will not come about just because someone has made up a song that is chanted to the strumming of a guitar. That may create an appropriate atmosphere for the change, but it will still require a great deal of scientific thought and engineering design and carefully supervised construction—and people with brains and training will have to do that.

Many are convinced that technology is at the root of our problems and claim that our complicated industrial apparatus must be dismantled and replaced by a way of life that is "closer to nature" and more econologically sound. But how can this be done in a world which contains more than four billion people and which never supported more than one billion in the days before industrialization?

If we grant that the antitechnology idealists don't want to see the death of three billion people, then we must suppose that as our present technology is dismantled, another one, simpler, less destructive, and even more efficient, must be simultaneously built up in order that the world's population continue to be supported. And that, too, requires scientific thought and engineering design and carefully supervised construction—and people with brains and training will have to do that.

It isn't hard to see, is it, that if we want things to work well, there is no substitute for brains and training?

And all we have to do is look about us to see that there isn't exactly an oversupply of brains and training.

Whatever direction Earth's history now takes; whether we are to opt for bigger and better technology, or smaller and better technology, we will need more brains and training than ever—that is, if we want civilization to survive, if we don't want it to collapse into an orgy of fighting and killing until its numbers shrink to the scattered few who can live by foraging and subsistence farming.

With brains and training that necessary, that *crucial,* isn't it a kind of will-to-suicide to brush off half the human race as a possible source of intelligence and will? Isn't it a kind of maximum stupidity to feel that we can solve the kind of problems we face by proceeding at half-steam?

Even at full-steam, we may not make it—but *half*-steam?

In other words, we need not only have more scientists and technologists than ever, but the very best we can find, wherever we can find them. By what supreme folly, then, do we assume that none of them are to be found among women? Why is it we so arrange our societies that half the human race rarely enters science or technology as a career, and that when some of that half manages to do so they find the path to better pay and leadership blocked step by the often unconscious but sometimes *expressed* view of the predominantly male subculture of science?

It is easy, of course, to sneer and say that women don't make good scientists, that science isn't women's work.

The whole concept of "women's work" is a fraud, however, since "women's work" is defined in a way that is convenient to men. If there are jobs men don't want to do and there is no handy minority to wish it on, it can always be given to women—the permanently available downtrodden majority.

As for science in particular not being women's work, it would be wearisome to go through the list of women who have contributed importantly to science, from Nobel laureates down—including some you may never have heard of, such as Voltaire's mistress, who was the first to translate Isaac Newton's *Principia Mathematica* into French (and did so with completely successful intelligence), and Lord Byron's daughter, who was one of the first two people to deal in detail with computer technology.

It might be argued that these women were exceptions (even exceptions "that prove the rule," to use an idiot phrase that depends on a misunderstanding of the meaning of the word "prove").

Of course, these are exceptions, but not because the vast bulk of women are not meant for science—only because the vast bulk of women can't get over the impossible hurdles placed in their way.

Imagine trying to become a scientist when you are constantly told you are not fitted for the task and are not smart enough; when most schools didn't let you enter; when the few schools that did didn't teach real science; when, if you managed to learn science somehow, the practitioners in the field met you with frozen aloofness or outright hostility and did their best to place you in a corner out of sight.

If that were a Black I was speaking of, any decent human being would be indignant over the situation and protest. But I'm speaking of a Woman, so many decent people look blank.

People who would hotly deny that there is any basic difference in intelligence among the "races" will still blandly believe that men are reasonable, logical, and scientific, while women are emotional, intuitive, and silly.

It isn't even possible to argue against this dichotomy sensibly, since the difference between the sexes is so taken for granted that it becomes self-fulfilling. From earliest childhood, we expect boys to act like boys, and girls to act like girls, and pressure each of them into it. Little boys are adjured not to be sissies and little girls are exhorted to be ladylike.

A certain indulgence can be permitted until adolescence, but woe betides the sissy and the tomboy thereafter. Once the classes in woodworking and home economics start, it is a tough girl who can insist on taking woodworking and an almost impossibly brave boy who can bear up under the universal execration that follows if he takes home economics.

If there is a natural division of aptitudes, why do we all work so hard to ridicule and prevent "exceptions"? Why not just let nature take its course? Do we know in our hearts that we have misread nature?

Once young people are old enough to grow interested in sex, the pressures of sex-differentiation become excruciating. Young men, having been well-indoctrinated into believing themselves the brainier sex, have the comfort of knowing that they are smarter than half the human beings in the world, however dumber they may be than other men. It would then be unbearable for a man to find a woman who showed herself to be smarter than he was. No charm, no level of beauty, would compensate.

Women don't need to find that out for themselves; they are nervously taught that by their mothers and older sisters. There's a whole world of training in the fine art of being silly and stupid—and attractive to boys who want to shine by contrast.

No girl ever lost a boy by giggling and saying, "Oh, please add up these figures for me. I can't add two and two for the life of me." She would lose him at once if she said, "You're getting it wrong, dear. Let me add that for you."

And no one can practice being silly and stupid long enough and hard enough without forgetting how to be anything else.

If you're a woman, you know what I'm talking about. If you're a man, find some woman who has no economic or social need to flatter you and ask her how hard she has to work at times never to seem smarter than her date.

Fortunately, I think these things are changing. They are not as bad as they were, say, a quarter-century ago. But there's still a long way to go.

With the world badly overpopulated, we no longer need numerous babies. In fact, we must have very few, no more than enough to replace the dying and, for a while, perhaps even fewer. This means we don't need women as baby-machines.

And if they are not to be baby-machines, they must have something else to do. If we want them to have but one or two babies at most, we must invite them out into the world, and make it worth their while to be there. We can't just give them menial, low-paying jobs as a matter of course. We must give them their fair chance at every possible branch of human endeavor on an equal basis with the male.

And most of all, *most* of all, women are needed in science. We cannot do without their brains. We cannot allow those brains to rest unused. We cannot, with criminal folly, destroy those brains deliberately, as we have been doing through all of history, on the plea that women must do "women's work" or, worse yet, that they must be "ladylike."

— 44 —

TV AND THE
RACE WITH DOOM

Mankind is plunging wildly toward catastrophe—and within a generation, per-haps. Many prophets tell us this in anguish. The world population is wildly rising, our environment is deteriorating, our cities decaying, our society disinte-grating, our quality of life declining.

How to halt it? How can we determine the necessary actions and then take them when the large majority of Earth's population is indifferent and is utterly concerned only with the immediate problem of the next meal. Or, if their eye is lifted to a further horizon, it is to become chiefly concerned with the hate and fear of some near neighbor.

To rally the different peoples of the world—different in language, religion, culture, and tradition—against the overriding problems that threaten to turn all the world into a desert, what weapon do we have?

The traditional weapon is force. Let one of the world's peoples, one coherent group, kill or conquer all the rest, make of itself the one power to be considered and let it then, out of its own wisdom and without having to consult or bargain with anyone else, take what measures are necessary to save the world!

But that won't do, for the nature of war at this time is such that even if we could bring ourselves to advocate world conquest, the very process of conquer-ing the world would destroy it and leave no room for any solution but death.

What else? As an alternative to force there is persuasion. Mankind must somehow be talked into saving itself, into agreeing to turn its combined strength and ingenuity toward a program for keeping the Earth fit for life.

But there is so little time, and mere persuasion has so rarely worked. Some new weapon of persuasion must be found.

Such a weapon exists. It is television.

Of all forms of communication, television is the most forceful and immedi-ate. It is not incomplete as radio and photography are; it is not remote, as books and printed periodicals are; it is not necessarily contrived fiction, as movies are. Television fills both eye and ear, in the full range of color and tone, and can deal with matters that are happening at the very moment they are being sensed.

Television has already shown its force. The Vietnam War was the first to be played out on television, and war has lost its glamor at least partly because it became distasteful when brought into our homes. The whole nation has been sharing experiences; every corner of the land is aware of the campus disturbances, of the drug scene, of the hippie counterculture. We all become, in a way, more neighborly since we share even in trivia, recognize the same ubiquitous faces, and become aware of the same advertising catch phrases.

And television is merely in its kindergarten stage. Its influence on society now only hints at what it could be when it comes of age.

There is a limit now to the number of television channels available in any community, a limit to the reach of any television station. The sharp limits leave room for little flexibility and imagination; only those programs that please a multitude are offered.

But communications satellites can change all that. By using objects in outer space as relays, signals can bounce from any one spot on Earth to any other, and the number of possible channels, stretching across a broad band of wavelengths, becomes virtually unlimited.

In 1965, the first commercial communications satellite, Early Bird, was launched. Its relay made available 240 voice circuits and one TV channel. In 1971 Intelsat IV will be in orbit with a capacity for 6,000 voice circuits and twelve TV channels. There is confident hope that such expansion of capacity will continue over the coming years at an even more rapid pace.

When sufficiently sophisticated satellites in sufficient numbers are placed in orbit, electronic communication for the first time will become personalized. Though television stations could still be offering mass programs to a mass audience, it will also be possible for any man on Earth, individually, to reach any other. With an unlimited number of voice and picture channels available, each man on Earth could dial any number on Earth—with or without visual accompaniment.

The printed word, in a computerized space-relay world, could be transmitted as efficiently as the spoken word. Facsimile mail, newspapers, magazines, and books could be readily available at the press of a button—anywhere.

The world will be tiny indeed, and we will all be neighbors, electronically. To be sure, we don't always love our neighbors; still, there is at least a greater chance of peace if we can talk easily among ourselves. It is simple to hate an abstraction, to cry down death upon some bunch of faceless foreigners somewhere. It is much harder to do the same to the pleasant foreigners with whom we could be speaking at any time—arguing our case and listening to theirs.

With massive personalized communication, the world will more and more share its problems and experiences. It will become less easy for any person to imagine that somehow his own corner of the world is divorced from the rest and that if he just minds his own business and hoes his own garden, the rest of the world can go to blazes.

As it happens, today we live in a world in which all parts will be dragged down to ruin if any major part is. None of us will ever escape if we don't all feel this unity-of-destiny with heart and soul; and television in its adulthood will make sure we feel this.

As communication improves and becomes more intensive, mankind will find it less necessary to live together in huge clusters. In a world of automation, it will be information that will have to be transported from point to point, not human bodies—and information can be moved electronically at the speed of light.

With unlimited numbers of television channels, conferences can meet in the form of images. The actual bodies belonging to those images, fed with facsimiles of any necessary documents, can be anywhere on Earth. Supervisors need not be at the automated factories in person, nor the average man at a concert. To a greater and greater extent, for more and more purposes, men can stay where they are and send their images.

This is not to say that people might not want to be together for personal or psychological reasons, or that they might not want to travel for fun and excitement. The point is that people will have less and less reason to travel when they *don't* want to.

The result will be that transportation facilities will feel lessened strain, and those who must still travel will be able to do so more comfortably and speedily. People will no longer have to live in groups so that they might reach each other. They can spread out.

In a world in which every person is, or can be, in instant touch with anyone else, we have in effect what has been called a "global village." Under such conditions, cities are unnecessary.

Again, this does not mean that some people won't choose to live together just because they want to. Still, the cities will decrease in size and become more livable for those who stay, while much of the population will spread out into the relatively empty areas, enjoying physical space *without* cultural isolation.

What about education? With a flexible electronic system in control, education can become infinitely more detailed and personal. The traditional school can remain for sports, physical culture, and the psychological values of social interaction. In addition, however, much of the educational process can be conducted at home, under individualized conditions, with an electronic tutor geared to the needs of each child. A happier and more creative generation will grow to adulthood.

Furthermore, education need not be confined only to the "advanced" nations. With mass electronics, the submerged mass of peasantry in Asia, Africa, and South America can, essentially for the first time, get the information it needs—information that the whole world needs to make sure peasantry gets.

The population of the have-not nations can grow up learning about modern agricultural methods in the most dramatic possible way—each for himself. They can learn the proper use of fertilizers and pesticides, proper hygiene, proper techniques for population control.

The beamed wavelengths, bouncing off satellites, can bypass clogged social setups, slip around decaying tradition, overcome the weight of illiteracy. The *whole* world, *all* of mankind, can receive the strong push into the approaching

twenty-first century that it must have. Nor must we underestimate the force of
the push. Radio, movies, and kindergarten television have already, in a mere
half-century, made astonishing progress toward Westernizing (even American-
izing) the world. What then will the communications revolution, grown to maturity, succeed in doing?

What of world government? Many consider it necessary if mankind is really
to take action against its overwhelming problems; yet world government seems
out of the question. There are too many conflicting interests, too many differences, too much hatred.

So it seemed when the infant American republic was born. The thirteen
original states were more spread-out in stagecoach days than the whole world is
now, and the distance between Boston and Savanna was as great in culture as in
miles. But canals, and then railroads, knit the country together, and a single
government became a practical possibility as well as a theoretical ideal.

Let mass electronic communication work, then, and become an intellectual
"railroad-net." Let the man on the banks of the Zambesi or the Orinoco have an
equal chance at information with the man on the banks of the Thames or the
Hudson, and there will come a morning when mankind will realize that for quite
awhile its governing units *had* been acting in a common cause, and that while
there were many nations, there was already the essence of world government.

Let's summarize: Television plus communications satellites will mean that,
for the first time in history, the planet can be a cultural unit. And that means
that for the first time in history it will have the capacity, and *perhaps* the will, to
be an action-unit against global problems. And maybe we will then survive.

Naturally, the communications revolution will not take place overnight. It
will not be here tomorrow. It will not be here, in full, by the year 2000.

But then mankind may not be destroyed overnight either.

What we face is a race. As the arteries of mass communication begin to
spread over the face of the earth, there will be a push toward efficient education,
decentralization, world unification and, on the whole, toward a stiffening of
action against the deteriorating situation.

In the same period, however, there is almost sure to be a continued rise in
population, a continued increase in pollution, a continued suicidal devotion of
man's efforts to dozens of separate and hostile military machines, and all this
will keep the situation deteriorating.

Which will win out?

My guess is that by the year 2000 the power of communication may still not
have reversed the tide, but misery will not have deepened as greatly as it would
have otherwise. Men will (we can hope) be more aware by then of the nature of
the race, of the terror of the doom, of the possibility of rescue. There will be a
greater push toward continuing the development of an intricate television network that will create the "global village."

If so, the forces of communication and unification may gain the upper hand
and—not without considerable pain and struggle—produce a new human civili-

zation that will look back upon the days we are living through now as a Dark Age.

On the other hand, it is also possible that the inertia of indifference, the dead weight of tradition, and the dark shadow of hate and suspicion may be too much for the developing communication network to overcome. In that case, we are probably doomed.

No man can, as yet, surely predict the outcome of the race, but each of us, whether we do it consciously or not, will pick his side. I am on the side of world communication, world understanding, and world union.

And you?

Afterword: The previous essay was written in 1971. Since then, we have had the experience of Ronald Reagan using television to spin his web of popularity. More surprisingly, the Soviets are learning to do that, and we have Mikhail Gorbachev smiling genially into the camera and presenting a charming wife to do the same. This is a revolution we ought to be more aware of.

THE NEXT SEVENTY YEARS
IN THE COURTS

I do not think there is some single inevitable future, graven on sapphire in the archives of Heaven, which humanity is fated to live out second by second through eternity. It is my opinion that there are only innumerable *possible* futures, any one of which may come to pass, depending on whatever it is that several billions of human beings decide to do from moment to moment right now, and depending also on outside circumstances that may be beyond the control of any of those billions.

From that point of view, the best one can do in trying to predict the future is to choose one that seems to be conceivable, given a reasonable choice of human motivations and responses together with a plausible development of technology—and one that seems interesting as well. I would like to present such a potential future based on two fundamental facts about the American legal system as it exists now.

First, the American legal system (unlike those in many other parts of the world) does not exist solely to serve the government. For two hundred years, it has recognized the importance of the individual, even when that individual seems to be in conflict with what are taken to be the interests of the government.

The accused is assumed to be innocent till proven guilty; he is guaranteed various liberties, such as those of speech, press, and religion; he is kept from unreasonable searches and seizures; he is assured of the right of trial by jury under detailed rules designed to keep the trial a fair one; and so on. Ideally, then, the American legal system takes seriously the notion of equal justice for all—rich and poor, powerful and powerless—without fear or favor.

Second, the American legal system does not nearly approach this ideal in actual fact. It is impossible for it to approach it as long as lawyers are not all equally skillful, as long as judges are not all immovable rocks of intelligent objectivity, and as long as a jury is no more intelligent and free of bias than the population from which it is chosen.

As we all know, rich men, corporations, and the government can hire the best lawyers and can afford to seize every opportunity for delay as they debate the issues endlessly. People of moderate income or less must hire such lawyers as they can afford, and even so the costs speedily become unbearable as delays continue. Whatever the merits of a poor man's case, he can be forced to the wall by nothing more than a waiting game. Justice can scarcely be served under such circumstances.

Again, judges and juries, being human, must have their quirks and biases, and the lawyer who happens to know best how to play upon these most skillfully is more apt to elicit favorable judgments than some other lawyer with nothing on his side but the better case.

In short, even in America the ideal of equal justice before the law tends to be a myth. Or, as Shakespeare has King Lear say:

> Through tattered clothes small vices do appear:
> Robes and furred gowns hide all. Plate sin with gold,
> And the strong lance of justice hurtless breaks;
> Arm it in rags, a pygmy's straw does pierce it.

We can summarize these two facts about the American legal system, then, by saying that it *doesn't* deliver justice, but that there is a strong belief that it *should*.

It follows, then, that Americans must ceaselessly attempt to have the actual legal system approach the ideal. Till now, that has seemed like an impractical dream. After all, lawyers and judges must differ among themselves in intelligence, skill, and integrity, while clients must differ among themselves in wealth, education, and social position. When people cannot help but be unequal, how can a system composed of people and working with people not be fatally influenced by this inequality?

Now, however, we are plunging headlong into a computerized society, and each year computers become more versatile and complex. Already, legal offices have computer outlets that can do library searches for applicable cases and precedents, and we can easily imagine that, as the decades pass, such searches can become more detailed.

Judges, too, will be capable of studying precedents in great detail, and we can imagine that an indispensable adjunct of the judge's bench will be a computer designed to advise on any decision that must be made. In many cases, the computer's decision, based on precedent, will be more knowledgeable and perhaps more just than any that might be come to without the thorough information stored in the computer's memory banks.

We might suppose that decisions reached on the basis of a computerized study of legal procedures and precedents, where both sides that stand before the bar have equal access to the computer and can see for themselves what the analysis of the situation is, would not be subjected to influence by differences in wealth or position among the contestants. Decisions "without fear or favor" will come within reach.

In fact, as computers make it possible to analyze cases from a legal position, in rapid detail, the whole functioning of law courts may be irretrievably revolutionized in a way that reminds me (in reverse) of warfare in fifteenth century Italy.

Italy was then divided into numerous city-states that were in a state of nearly perpetual warfare among themselves. They hired mercenary armies that did the fighting in return for generous payment, while the merchants and artisans of the cities lived in peace. The mercenary captains were anxious to earn their pay but did not want to lose their armies, so the wars came to be largely ones of maneuver. When one captain saw that he was in a position of such disadvantage that he would surely lose if he fought, he didn't fight but gave in and marched away. He might have better luck next time.

This worked well until 1494, when a French army marched into Italy. The French did not fight by the rules. They fought at night; they fought in winter; they didn't worry about casualties; they didn't march away if outmaneuvered. The Italian armies were pushed aside, and for three and a half centuries Italy was a battleground for foreigners: French, Spanish, and German. Italy's glorious Renaissance civilization was destroyed, and warfare became completely uncivilized, reaching a bitter climax during the wars of religion a century and a half later.

The law courts are in the uncivilized stage now. Competing lawyers need not consider results as inevitable; they can wait for the opposition to make mistakes; they can hope for a judge to make a blunder or for a jury to be hoodwinked. Virtually every case, therefore, is worth trying, for it offers at least some hope for victory. The result is that the United States of the 1980s is perhaps the most litigious society the world has ever seen.

With computers in action, there will be increasingly little chance of an egregious mistake in legal strategy or tactics, and a diminishing hope of unpredictability on the part of the judge. Lawyers on either side would be able to predict, with far greater assurance than today, just how a case will go, whether a client will be judged guilty or not, whether damages will have to be paid and how much. Both sides will probably have much the same information, and each will realize that one side or the other is at a distinct disadvantage. Like the mercenary armies of fifteenth century Italy, the most economical strategy will then be a quiet retreat for the side that faces likely defeat.

One can see the consequences. As computers grow more elaborate and versatile and can better cope with the complexities of law, judgments, and precedents, there will be fewer appeals, fewer strategies of delay. There will be faster and shorter trials, more settlements out of court, and, most of all, fewer cases brought to trial in the first place. In fact, we might go even further and suggest that there will be fewer cases of cutting corners or playing games with the law because there will be less chance of hoodwinking an opposing council, or a judge or jury. The litigiousness of the American public will shrink rapidly in the course of the next seventy years in that case, and the habit of obeying the law would take a much firmer hold.

Of course, as we consider the whole concept of computerized law, we may experience a sense of horror. Won't we be removing humanity from the courtroom? Won't we be subjecting questions of justice, of punishment and mercy, to a cold-blooded machine?

Yes, but so what?

In the first place, a machine would have to be cold-blooded indeed to match the cold-bloodedness of human beings in courts in which justice is *not* even-handed between the rich and the poor, between the educated and the unedu-cated, between those who are socially acceptable and those who are "different."

In the second place, we are already dependent on machines. Who would not accept a breathalyzer as a judge of drunkenness over the personal estimate of a policeman? Who would not accept careful forensic medical analysis over "eye-witness" evidence? And if we develop computerized methods for judging which of two contradictory witnesses is more likely to be lying, would we not feel more confidence in that than in the decision of a jury which might be swayed by the fact that one witness is clean-cut and good-looking and the other is not?

It is possible, of course, that lawyers, judges, and all the other personnel of the American judicial system might feel a little anxious over their personal fates. In a nonlitigious society, what will there be for lawyers and the rest to do? Will an increasing number of them be piloting taxicabs and waiting on tables in the course of the next seventy years?

Not necessarily. In fact, it might well be argued that there will be *more* work for lawyers, not less, as time goes on; and better work, too. There will be work that is more useful and more important.

Where casework is concerned, lawyers will have to make careful use of the computer to decide the likely outcome and will have to advise their clients accordingly. They would earn (and deserve) good payment for that.

Then, too, computers capable of playing an important role in the legal system do not come into existence just like that; they have to be programmed. If the computerization of the law is to make sense, the programming must be done by experts, those who understand the details and purpose of the law and who also understand the workings of the computer well enough to make sure that those details and that purpose will be firmly upheld. And who can do that but lawyers, and *good* lawyers at that?

Finally, think of the new laws that will be needed over the next seven decades. The whole legal system will have to be overhauled in order to be better adapted to computerization. It will have to be more carefully quantified. Ex-tenuating circumstances will have to be more specifically defined, and so on.

There will be wholly new crimes the law will have to consider. We already have to deal with computer-crime. Such wrongdoing will gain whole new di-mensions of horror if it becomes possible for unscrupulous individuals to tamper with the very computers that deal with crime. In fact, so dependent will society become on computers in the coming decades, quite apart from its pos-sible role in connection with the law, that tampering with computers in any way may well become looked upon as the most dreadful and unforgiveable of all crimes—for such crime will tend to destroy the very basis of society. How do we deal with that?

Then, too, how do we deal with the "rights" of thinking machines? At what point will robots have to be considered as more than "mere" machines? If it is increasingly recognized that animals have rights and may not be treated with deliberate cruelty, might there not be a similar feeling for machines that seem more intelligent than any animal?

And what of the new realms of human occupancy that will be opened up in the next seven decades? The "law of the sea" will be increasingly pressing as we learn to mine the seafloor, and as we begin increasingly to open the continental shelves to human occupancy.

Again, "space law" will have to be written in great detail as more and more human structures—space stations, solar power stations, observatories, laboratories, factories, orbiting cities—are placed in space.

Come to think of it, there will be so much important, difficult, and novel work for lawyers to do in the next seventy years that they, more than anyone else, should push for a decrease in the petty and useless casework that clogs the courts and hardens the arteries of the legal mind. They should welcome computerization with glad cries, if that will do it.

46

THE FUTURE OF COSTUME

We live in an era of futurism. Until about 1800 changes in society took place so slowly in time, and spread so slowly through space, that people generally took it for granted that there was no change.

Change, however, has accelerated throughout history, and by 1800, with the Industrial Revolution in full swing, it became fast enough to notice. What's more, change has continued to take place and more and more quickly until it is now a whirlwind that no one can ignore.

Any significant change in society is bound to produce a change in costume as well.

For instance, we now live in an age of growing egalitarianism. Class distinctions, whether due to birth, position, or money, are fading, at least as far as appearance is concerned, and this has led to a simplification of clothing.

Complexity of costume, after all, is always an important way of indicating that a particular person is upper class. The costume must be too expensive for a lower-class person, it must be impractical enough to indicate that the wearer does not have to engage in physical work, and it must be complicated enough to indicate that the wearer requires the help of a valet or maid.

We have watched such complexities disappear. Top hats have disappeared except in show business. So have derbies, straw hats, and hatware in general, except when they are needed for warmth, military display, or for some archaic ritual such as an Easter parade. In the same way, men's vests have disappeared, as have cuffs on the trousers and watchpockets. Ties may yet disappear, and so may buttons.

This has been even more marked in the case of women, where the elaborate clothing of the upper-class did everything possible to emphasize that the feminine role was purely decorative. Since World War II, as women have entered the work force and the feminist movement has become more pronounced, women's clothing has of necessity become simpler and more free.

The rapidity of change now means that clothing designers, if they are to be successful, must think of where that change is taking us. For instance, the trend seems to be for increasing equality of the sexes in the workplace and social relationships. There has therefore been a continuing sexual convergence of costume.

Since in the past the male human being has been accepted as the "superior" sex, female garb has been (unconsciously, at least) a badge of inferiority. When women began to put on pants or wear shirts, they felt daring and bold. Men do not show an equal tendency to wear skirts or carry purses. In order for them to do so they must overcome a sense of "femininity" (that is, inferiority).

The increasing tendency toward unisex costuming, then, will continue in the direction of male wear for both men and women. Despite this, the designer can, of course, expend infinite ingenuity in making certain that, however similar the costumes, there will remain unmistakable ways of distinguishing the sexes. The subtlety involved will not only exercise the artistry of the designer in a satisfying and creative way, but will make the distinction more interesting and stimulating for being less blatant.

Looking farther into the future, we may see the end of commuting. Electronic communications may make it possible to conduct much administrative and supervisory work from the home. This may further increase informality and may even totally blur the distinction between costume appropriate for the home and for the office.

It may become common to wear clothing that has the cut and appearance of pajamas and bathrobes for use in public (as in conferences by closed-circuit TV) or even in visiting friends. Naturally, such clothing would be designed with that fact in mind and it would be nattier, more attractive, and more substantial than what we now wear for lounging about the house—but just as easy to get in and out of.

In particular, footwear will become less formal. The tight leather box of men's shoes will open up and the tall heel of women's shoes will come down. Sandals and slippers will become proper wear, or at least shoes that will appear to be sandals and slippers.

If we continue to look into the future, we will have to adjust our sights to something wider and higher than the earth. There may well come a time in the course of the twenty-first century when we will be living through the beginnings of a space-centered society. There may be space settlements, each of which may house ten thousand human beings or more.

Such settlements will not have climate as we know it. There will be nothing out of control, nothing unpredictable (except for a very rare strike by a pebble-sized meteoroid, or by unusually energetic cosmic ray particles). We can have late spring weather if we want it, or an eternal summer, or a brisk fall day on order. Nor need we expect pollution or unusual quantities of soot or dirt.

It may well be that on most (or perhaps all) such settlements, temperature extremes will simply not be found, nor any violence of wind or storm. In such cases, clothes might not be needed for warmth or protection. They might become pure ornament-plus-comfort.

In such cases, individualism might reach an extreme. The nudity taboo might vanish totally in some settlements, but it is more likely that clothes will remain, if only because a properly clothed body is more attractive (in all but a minority of cases) than our natural skin and hair.

Each person's body would then become a palette upon which clothing—

color, form, material—would be so chosen as to match a personality, an occasion, or a mood.

And the clothing designer will then find himself an acknowledged exponent of one of the fine arts—a painter in fabric, beautifying anatomy.

47

THE IMMORTAL WORD

You may have heard the statement: "One picture is worth a thousand words."

If so, don't you believe it. It may be true on occasion—as when someone is illiterate, or when you are trying to describe the physical appearance of a complicated object. In other cases, the statement is nonsense.

Consider, for instance, Hamlet's great soliloquy that begins with "To be or not to be," that poetic consideration of the pros and cons of suicide. It is 260 words long. Can you get across the essence of Hamlet's thought in a quarter of a picture—or, for that matter, in 260 pictures? Of course not. The pictures may be dramatic illustrations of the soliloquy if you already know the words. The pictures by themselves, to someone who has never read or heard *Hamlet*, will mean nothing.

As soon as it becomes necessary to deal with emotions, ideas, fancies— abstractions in general—only words will suit. The modulation of sound, in countless different ways, is the only device ever invented by human beings that can even begin to express the enormous complexity and versatility of human thought.

Nor is this likely to change in the future. You have heard that we live in an "age of communication," and you may assume, quite rightly, that amazing and fundamental changes are taking place in that connection. These changes, how- ever, involve the *transmission* of information, not its nature. The information itself remains in the form it was in prehistoric times: speech, and the frozen symbology of speech that we call writing.

We can transmit information in sign language, by semaphor, by blinking lights, by Morse code, by telephone, by electronic devices, by laser beams, or by techniques yet unborn—and in every case what we are doing is trans- mitting words.

Pictures will not do; they will never do. Television is fun to watch, but it is entirely dependent on the spoken and written word. The proof is this: darken the image into invisibility but leave the sound on, and you will still have a crude sense of what is going on. Turn off the sound, however, and exclude the appear- ance of written words, and though you leave the image as bright as ever, you will find you understand nothing of what is going on unless you are watching the most mindless slapstick. To put it even more simply: radio had no images at all and managed, but the silent movies found subtitles essential.

There is the fundamental rule, then. In the beginning was the word (as the Gospel of St. John says in a different connection), and in the end will be the word. The word is immortal. And it follows from this that just as we had the writer as soon as writing was invented five thousand years ago, so we will have the writer, of necessity, for as long as civilization continues to exist. He may write with other tools and in different forms, but he will *write*.

Having come to the conclusion that writers have a future, we might fairly ask next: What will the role of the writer be in the future? Will writers grow less important and play a smaller role in society, or will they hold their own?

Neither.

It is quite certain that writers' skills will become steadily more important as the future progresses—providing, that is, that we do not destroy ourselves, and that there *is* a future of significance, one in which social structures continue to gain in complexity and technological advance.

The reasons are not difficult to state.

To begin with, technological advance has existed as long as human beings have. Our hominid ancestors began to make and use tools of increasing complexity before the present-day hominid we call *Homo sapiens* had yet evolved. Society changed enormously as technology advanced. Think what it meant to human beings when agriculture was invented—then herding, pottery, weaving, and metallurgy. Then, in historic times, think of the changes introduced by gunpowder—the magnetic compass, printing, the steam engine, the airplane, and television.

Technological change feeds on previous technological change, and the rate of change increases steadily. In ancient times, inventions came so infrequently that individual human beings could afford to ignore them. In one person's generation, nothing seemed to change as far as the social structure and quality of life was concerned. But as the rate of change increased, that became less true, and after 1800 the Industrial Revolution made it clear that life—everyday life—was changing rapidly from decade to decade and then from year to year and, by the closing portion of the twentieth century, almost from day to day. The gentle zephyr of change that our ancestors knew has become a hurricane.

We know that change is a confusing and unsettling matter. It is difficult for human beings to adjust to change. There is an automatic resistance to change, and that resistance diminishes the advantages we can obtain from change. From generation to generation, then, it has become more and more important to explain the essentials of change to the general public, making it aware of the benefits—and resultant dangers—that are derived from change. That has never been more important than it is now; and it will be steadily more important in the future.

Since almost all significant change is the result, directly or indirectly, of advances in science and technology, what we're saying is that one particular type of writing—writing about science—will increase in importance even more quickly than writing in general will.

We live in a time when advances in science and technology can solve the problems that beset us by increasing the food supply, placing reproductive potentialities under control, removing pollution, multiplying efficiency, obtaining new sources of energy and materials, defeating disease, expanding the human range into space, and so on.

Advances in science and technology also create problems to bedevil us: producing more dangerous weapons, manufacturing increasingly insidious forms of pollution, destroying the wilderness, and disrupting the ecological balance of Earth's living things.

At every moment, the politicians, the businessmen, and to some extent every portion of the population must make decisions on both individual and public policy that will deal with matters of science and technology.

To choose the proper policies, to adopt this and reject that, one must know something about science and technology. This does not mean that everyone must be a scientist, as we can readily see from the analogy of professional sport and its audience. Millions of Americans watch with fascinated intentness games of baseball, football, basketball, and so on. Very few of them can play the game with any skill; very few know enough to be able to coach a team; but almost all of them know enough about the game to appreciate what is going on, to cheer and groan at appropriate times, and to feel all the excitement and thrills of the changing tides of fortune. That must be so, for without such understanding watching a game is merely a matter of watching chaos.

And so it must be that as many people as possible must know enough about science and technology to at least be members of an intelligent *audience*.

It will be the writer, using words (with the aid of illustrations where they can make the explanation simpler or raise the interest higher), who will endeavor to translate the specialized vocabulary of science and technology into ordinary English.

No one suggests that writing about science will turn the entire world into an intelligent audience, or that such writing will mold the average person into a model of judgment and creative thought. It will be enough if this writing spreads scientific knowledge as widely as possible; if some millions, who would otherwise be ignorant (or, worse, swayed by meaningless slogans), would gain some understanding as a result; and if those whose opinions are most likely to be turned into action, such as the political and economic rulers of the world, are educated.

H. G. Wells said that history was a race between education and catastrophe, and it may be that the writer will add just sufficient impetus to education to enable it to outrace catastrophe. And if education wins by even the narrowest of margins, how much more can we ask for?

Nor is a world that is oriented more in the direction of science and technology needed merely for producing better judgments, decisions, and policies. The very existence of science and technology depends upon a population that is both understanding and sympathetic.

There was a time when science and technology depended strictly on individual ideas, individual labor, and individual financial resources. We are terribly attracted to the outmoded stereotype of the inventor working in his home workshop, of the eccentric scientist working in his home laboratory, of the universe of ignorance being assaulted by devices built of scraps, string, and paste.

It is so no longer. The growing complexity of science and technology has outstripped the capacity of the individual. We now have research teams, international conferences, industrial laboratories, large universities. And all this is strained, too.

Increasingly, the only source from which modern science and technology can find sufficient support to carry on its work is from that hugest repository of negotiable wealth—the government. That means the collective pocketbooks of the taxpayers of the nation.

There never has been a popular tax or an unreluctant taxpayer, but some things will be paid for more readily than others. Taxpayers of any nation are usually ready to pay enormous sums for military expenses, since all governments are very good at rousing hatred and suspicions against foreigners.

But an efficient military machine depends to a large extent on advances in science and technology, as do other more constructive and less shameful aspects of society. If writers can be as effective in spreading the word about science and technology as governments are at sowing hatred and suspicion, then public support for science is less likely to fail, and science is less likely to wither.

Moreover, science and technology cannot be carried on without a steady supply of scientists and engineers, an increasing supply as the years go on. Where will they come from?

They will come from the general population, of course. There are some people who gain an interest in science and technology in youth and can't be stopped, but they, by themselves, are simply not numerous enough to meet the needs of the present, let alone the future. There is a much larger number of youngsters who would gain such an interest if they were properly stimulated, but perhaps not otherwise.

Again, it is the writer who might catch the imagination of young people, and plant a seed that will flower and come to fruition. I have received a considerable number of letters from scientists and engineers who have taken the trouble to tell me that my books turned them toward science and technology. I am quite convinced that other science writers get such letters in equal numbers.

Let me make two points, however.

First, in order to write about science, it is not entirely necessary to be deeply learned in every aspect of science (no one can be, these days) or even in one aspect—although that helps. To know science well can make you a "science writer," but any intelligent person who has a good layman's acquaintance with the scientific and technological scene can write a useful article on some subject related to science and technology. He can be a *writer* dealing with science.

Here is an example of what I have in mind.

Digital clocks seem to be becoming ever more common these days, and the old fashioned clock dial seems to be fading away. Does that matter? Isn't a digital clock more advanced? Doesn't it give you the time more accurately? Won't children be able to tell time as soon as they can read instead of having to learn how to decipher the dial?

Yet there are disadvantages to a possible disappearance of the dial that perhaps we ought to keep in mind.

There are two ways in which anything might turn—a key in a lock, a screw in a piece of wood, a horse going around a race track, Earth spinning on its axis. They are described as "clockwise" and "counterclockwise." The first is the direction in which the hands on a clock move; the second is the opposite direction. We are so accustomed to dials that we understand clockwise and counterclockwise at once and do not make a mistake.

If the dial disappears (and, of course, it may not, for fashion is unpredictable) the terms clockwise and counterclockwise will become meaningless—and we have no adequate substitutes. If you clench your hands and point the thumbs upward, the fingers of the left hand curl clockwise and those of the right hand counterclockwise. You might substitute "left-hand twist" and "right-hand twist," but no one stares at their clenched hands as intently and as often as at clock dials, and the new terms will never be as useful.

Again, in looking at the sky, or through a microscope, or at any view that lacks easily recognizable reference marks, it is common to locate something by the clock dial. "Look at that object at eleven o'clock," you may say—or five o'clock, or two o'clock, or whatever. Everyone knows the location of any number from one to twelve on the clock dial and can use such references easily.

If the dial disappears, there will again be no adequate substitute. You can use directions, to be sure—northeast, south-by-west, and so on, but no one knows the compass as well as the clock.

Then, too, digital clocks can be misleading. Time given as 5:50 may seem roughly five o'clock, but anyone looking at a dial will see that it is nearly six o'clock. Besides, digital clocks only go up to 5:59 and then move directly to 6:00, and youngsters may be confused as to what happened to 5:60 through 5:99. Dials give us no such trouble.

One can go on and find other useful qualities in dials versus digits, but I think the point is clear. An article can be written that has meaning as far as technology is concerned and will provoke thought and yet not require a specialist's knowledge. We can't all be science writers, but we can all be writers about science.

The second point to be made is that I do *not* say that writers won't be needed in increasing numbers in other fields.

As computers and robots take over more of the dull labor of humanity and leave human beings to involve themselves in more creative endeavors, education will have to change in such a way as to place increasing emphasis on creativity. No doubt, education by computer will become more and more important, and a new kind of writer—the writers of computer programs for education—will arise and become important.

Again, as leisure time continues to increase the world over, writing in the form of books, plays, television, movie scripts, and so on, will be needed in greater numbers to fill that leisure time.

In other words, more and more writers of more and more different kinds will be needed as time goes on; but of them all it is writers on science for whom the need will grow most quickly.

48

LIBERTY IN THE
NEXT CENTURY

For a hundred years "Liberty Enlightening the World" (better known as "The Statue of Liberty") has stood in New York Harbor, and the famous sonnet on the plaque on its base reads, in part:

> . . . *Give me your tired, your poor,*
> *Your huddled masses yearning to breathe free.* . .

It's the phrase "huddled masses" that seems to me to be the key. A crowded world, lacking liberty, was urged to send its excess population to the greater freedom of the relatively empty spaces of the United States.

This makes sense, for crowds are the enemies of liberty. Despite all traditional, philosophic, and legal supports to the idea of freedom, it still remains, as the saying goes, that "the freedom of your fist ends where my nose begins." The more people you are surrounded by, the more surely your fist, in its aimless wanderings, will make contact with someone's nose; and, therefore, the more limited is your freedom.

In reverse, history is full of peoples who emigrated to emptier lands in search of freedom. Russian serfs fled to the Ukraine to join the free Cossack bands; Europeans went to North America and Australia. Americans from the crowded east steadily drifted westward.

In 1886, however, even as the Statue of Liberty was being dedicated, the time of empty spaces was already coming to an end, and the pressure of population was continuing to build. The population of the world was 1.7 billion in 1886, and it is 4.90 billion now, an increase of 280 percent. The population of the United States was 57 million in 1886 and it is 240 million now, an increase of 420 percent.

Does that really limit freedom? Yes, of course it does. No matter how much liberty might be worshipped and maintained in the abstract, it will be increasingly blocked in the performance.

If two people live in an apartment with two full bathrooms, each may use a bathroom at will, whenever and for however long he or she wishes. If twenty people live in an apartment with two full bathrooms, a system of rationing must be set up, and no individual will possibly be able to use a bathroom at will. Philosophic beliefs in "freedom of the bathroom" and constitutional guarantees of the same will not help. And if that seems purely hypothetical to you, try to get a taxi during rush hour by telling yourself that you have a *right* to hail a taxi when you wish.

Well, then, what will happen in the course of the coming century, during which (barring such disasters as thermonuclear war or the coming of deadly new plagues) population is bound to increase still further. Must we look forward to an inevitable withering of liberty?

Not entirely, perhaps. To a certain limited extent, it may be that a new frontier will open, and that there will therefore be the beginning of a vista of a new increase in liberty.

The twenty-first century may see humanity spread outward into space. Not only could there be solar power stations in space, observatories, laboratories, and factories—all built out of materials made available by mining stations set up on the Moon—but there may even be settlements in high orbit about the Earth.

We can easily envisage artificial structures capable of holding ten thousand human beings, or perhaps more, without unreasonable crowding. We could not possibly build them fast enough to absorb the coming population increase on Earth, but they will represent a beginning, and Earth can continue to seek out benign and humane ways of limiting population growth by cutting the birth rate.

The settlements can be so arranged as to have sunlight enter or be blocked. They can set up an artificial day-night alternation. By having the settlements rotate, an artificial gravity-like effect may be induced. The interior will have farmland, buildings, an absolutely benign weather pattern, a balanced ecology excluding undesirable weeds, parasites, etc. In short, there would be the possibility, at least, of idyllic surroundings.

The building of such space settlements offers a third historic opportunity for increasing the liberty and variety of the human species. In the eighth century B.C., Greeks and Phoenicians settled along the shores of the Mediterranean and Black Seas, founding new cities. In the seventeenth century A.D., European settlers founded new cities along the western shores of the Atlantic. And now in the twenty-first century, new cities will be founded in space.

We might visualize dozens of such cities in the next century, and eventually hundreds, and even thousands, as humanity reaches out with new technologies for the asteroid belt and converts asteroid after asteroid into homes for people.

Those who first populate one of these settlements will undoubtedly set up the social and economic conventions under which they will exist. They will continue to live under those laws that they are accustomed to and that they approve, but they may well adapt, renovate, reform, or totally invent other laws that, they may think, will represent improvements.

We needn't assume that every settlement will choose a way of life that represents some sort of American ideal. Not every settlement will be a picture of pastoral small-town America. Different groups of people from different nations will be involved in such settlements. It may be that there will be a space settlement which is Shiite in nature, or Fundamentalist Baptist, or Orthodox Jewish, or Communist, or some things we cannot easily imagine.

While such specialized settlements may have freedom in their sense, they may not have freedom in our sense. Thus, the Puritans in the early seventeenth century came to Massachusetts in order to escape religious persecution and be "free." As a result, they are often considered as heroes of the movement for increasing freedom. The fact is, however, that they set up a theocratic state in which *only they* could be free, and where people who didn't agree with them were persecuted, driven out, and in a few cases even hanged.

Not all such exclusivistic experiments will succeed, however, any more than the illiberal Puritans of Massachusetts could forever prevent people from thinking for themselves or establishing less tightly controlled communities. Then, too, there are bound to be eclectic settlements established as well, those in which people can think for themselves and in which ways of life will not be strictly encoded.

If that is so, it may be that people will "vote with their ships." Those who find life too limited and straitened in one settlement may emigrate to another where freedom in a wider sense flourishes. Those settlements that offer a better life on the whole will grow at the expense of those that do not, and the test of survival will perhaps see to it that liberty increases.

Of course, population will continue to increase even in individual space settlements, and such an increase will far more quickly make its adverse side-effects visible in the limited space of a space settlement than on the vast surface of Earth.

Two courses of action will then be possible—more space settlements must be built into which the excess population of an older settlement can be transferred, or there must be a careful and effective limiting of the birth rate.

In the long run, the number of additional space settlements cannot be made to match a population that grows exponentially. It is quite easy for populations to double every thirty years, but it will be increasingly difficult for even the busiest space builders to double the total number of space settlements every thirty years. Consequently, the limitation of population through a strictly controlled birth rate will eventually force itself on Earth and space settlements alike, and this in itself is something that places a limit on total liberty.

But then, total liberty has always been an unrealizable abstraction.

It seems to me as we envisage the beginning of this movement into space in the next century, that certain consequences are clear.

There are bound to be settlements that will not approve of other settlements and that will try to limit emigration, immigration, and even cultural contact. There will be the drive to keep one's own way of life unspoiled by the invasion of strange ideas.

Even those settlements that welcome new people and new ideas may not welcome the strange parasites that people from other settlements (and most particularly, from Earth) will bring. It may be that the ecological balance on any one settlement will have to be protected from the upset induced by intrusion from beyond.

There will thus be a continuing impulse for individual settlements, making use of advanced technologies, to leave the solar system forever and throw themselves upon the Universe as small but independent worlds.

This will not be altogether bad, for it may be the way in which humanity will come to explore the Galaxy, and the Universe generally, finding other planetary systems and populating them. In all this, many millions of years will pass. Even if human beings don't encounter other intelligences, evolutionary forces will convert humanity into innumerable related species—all intelligent, but perhaps intelligent in different ways.

And here my imagination begins to grapple with more than it can grasp, and I call a halt.

49

THE VILLAIN
IN THE ATMOSPHERE

The villain in the atmosphere is carbon dioxide.

It does not seem to be a villain. It is not very poisonous and it is present in the atmosphere in so small a quantity that it does us no harm. For every 1,000,000 cubic feet of air there are only 340 cubic feet of carbon dioxide—only 0.034 percent.

What's more, that small quantity of carbon dioxide in the air is essential to life. Plants absorb carbon dioxide and convert it into their own tissues, which serve as the basic food supply for all of animal life (including human beings, of course). In the process, they liberate oxygen, which is also necessary for all animal life.

But here is what this apparently harmless and certainly essential gas is doing to us:

The sea level is rising very slowly from year to year. The high tides tend to be progressively higher, even in quiet weather, and storms batter at breakwaters more and more effectively, erode the beaches more savagely, batter houses farther inland.

In all likelihood, the sea level will continue to rise and do so at a greater rate in the course of the next hundred years. This means that the line separating ocean from land will retreat inland everywhere. It will do so only slightly where high land abuts the ocean. In those places, however, where there are low-lying coastal areas (where a large fraction of humanity lives) the water will advance steadily and inexorably and people will have to retreat inland.

Virtually all of Long Island will become part of the shallow offshore sea bottom, leaving only a line of small islands running east to west, marking off what had been the island's highest points. Eventually the sea will reach a maximum of two hundred feet above the present water level, and will be splashing against the windows along the twentieth floors of Manhattan's skyscrapers. Naturally the Manhattan streets will be deep under water, as will the New Jersey shoreline and all of Delaware. Florida, too, will be gone, as will much of the British Isles, the northwestern European coast, the crowded Nile valley, and the low-lying areas of China, India, and the Soviet Union.

It is not only that people will be forced to retreat by the millions and that many cities will be drowned, but much of the most productive farming areas of the world will be lost. Although the change will not be overnight, and though people will have time to leave and carry with them such of their belongings as they can, there will not be room in the continental interiors for all of them. As the food supply plummets with the ruin of farming areas, starvation will be rampant and the structure of society may collapse under the unbearable pressures.

And all because of carbon dioxide. But how does that come about? What is the connection?

It begins with sunlight, to which the various gases of the atmosphere (including carbon dioxide) are transparent. Sunlight, striking the top of the atmosphere, travels right through miles of it to reach the Earth's surface, where it is absorbed. In this way, the Earth is warmed.

The Earth's surface doesn't get too hot, because at night the Earth's heat radiates into space in the form of infrared radiation. As the Earth gains heat by day and loses it by night, it maintains an overall temperature balance to which Earthly life is well-adapted.

However, the atmosphere is not quite as transparent to infrared radiation as it is to visible light. Carbon dioxide in particular tends to be opaque to that radiation. Less heat is lost at night, for that reason, than would be lost if carbon dioxide were not present in the atmosphere. Without the small quantity of that gas present, the Earth would be distinctly cooler on the whole, perhaps a bit uncomfortably cool.

This is called the "greenhouse effect" of carbon dioxide. It is so called because the glass of greenhouses lets sunshine in but prevents the loss of heat. For that reason it is warm inside a greenhouse on sunny days even when the temperature is low.

We can be thankful that carbon dioxide is keeping us comfortably warm, but the concentration of carbon dioxide in the atmosphere is going up steadily and that is where the villainy comes in. In 1958, when the carbon dioxide of the atmosphere first began to be measured carefully, it made up only 0.0316 percent of the atmosphere. Each year since, the concentration has crept upward and it now stands at 0.0340 percent. It is estimated that by 2020 the concentration will be about 0.0660 percent, or nearly twice what it is now.

This means that in the coming decades, Earth's average temperature will go up slightly. Winters will grow a bit milder on the average and summers a bit hotter. That may not seem frightening. Milder winters don't seem bad, and as for hotter summers, we can just run our air-conditioners a bit more.

But consider this: If winters in general grow milder, less snow will fall during the cold season. If summers in general grow hotter, more snow will melt during the warm season. That means that, little by little, the snow line will move away from the equator and toward the poles. The glaciers will retreat, the mountain tops will grow more bare, and the polar ice caps will begin to melt.

That might be annoying to skiers and to other devotees of winter sports, but would it necessarily bother the rest of us? After all, if the snow line moves north, it might be possible to grow more food in Canada, Scandinavia, the Soviet Union, and Patagonia.

Still, if the cold weather moves poleward, then so do the storm belts. The desert regions that now exist in subtropical areas will greatly expand, and fertile land gained in the north will be lost in the south. More may be lost than gained.

It is the melting of the ice caps, though, that is the worst change. It is this which demonstrates the villainy of carbon dioxide.

Something like 90 percent of the ice in the world is to be found in the huge Antarctica ice cap, and another 8 percent is in the Greenland ice cap. In both places the ice is piled miles high. If these ice caps begin to melt, the water that forms won't stay in place. It will drip down into the ocean and slowly the sea level will rise, with the results that I have already described.

Even worse might be in store, for a rising temperature would manage to release a little of the carbon dioxide that is tied up in vast quantities of limestone that exist in the Earth's crust. It will also liberate some of the carbon dioxide dissolved in the ocean. With still more carbon dioxide, the temperature of the Earth will creep upward a little more and release still more carbon dioxide.

All this is called the "runaway greenhouse effect," and it may eventually make Earth an uninhabitable planet.

But, as you can see, it is not carbon dioxide in itself that is the source of the trouble; it is the fact that the carbon dioxide concentration in the atmosphere is steadily rising and seems to be doomed to continue rising. Why is that?

To blame are two factors. First of all, in the last few centuries, first coal, then oil and natural gas, have been burned for energy at a rapidly increasing rate. The carbon contained in these fuels, which has been safely buried underground for many millions of years, is now being burned to carbon dioxide and poured into the atmosphere at a rate of many tons per day.

Some of that additional carbon dioxide may be absorbed by the soil or by the ocean, and some might be consumed by plant life, but the fact is that a considerable fraction of it remains in the atmosphere. It must, for the carbon dioxide content of the atmosphere is going up year by year.

To make matters worse, Earth's forests have been disappearing, slowly at first, but in the last couple of centuries quite rapidly. Right now it is disappearing at the rate of sixty-four acres per minute.

Whatever replaces the forest—grasslands or farms or scrub—produces plants that do not consume carbon dioxide at a rate equal to that of forest. Thus, not only is more carbon dioxide being added to the atmosphere through the burning of fuel, but as the forests disappear, less carbon dioxide is being subtracted from the atmosphere by plants.

But this gives us a new perspective on the matter. The carbon dioxide is not rising by itself. It is people who are burning the coal, oil, and gas, because of their need for energy. It is people who are cutting down the forests, because of their need for farmland. And the two are connected, for the burning of coal and oil is producing acid rain which helps destroy the forests. It is *people,* then, who are the villains.

What is to be done?

First, we must save our forests, and even replant them. From forests, properly conserved, we get wood, chemicals, soil retention, ecological health—and a slowdown of carbon dioxide increase.

Second, we must have new sources of fuel. There are, after all, fuels that do not involve the production of carbon dioxide. Nuclear fission is one of them, and if that is deemed too dangerous for other reasons, there is the forthcoming nuclear fusion, which may be safer. There is also the energy of waves, tides, wind, and the Earth's interior heat. Most of all, there is the direct use of solar energy.

All of this will take time, work, and money, to be sure, but all that time, work, and money will be invested in order to save our civilization and our planet itself.

After all, humanity seems to be willing to spend *more* time, work, and money in order to support competing military machines that can only destroy us all. Should we begrudge *less* time, work, and money in order to save us all?

50

THE NEW LEARNING

Suppose you buy a computer that can play chess with you. You can punch in your move and it will indicate a countermove.

It may not occur to you, when you do so, that you are holding in your hand something that could symbolize a greater change in society than anything the steam engine was responsible for.

Follow it through.

You like to play chess. You're not very good at it, but you enjoy it. With the computer you can play chess and have a little fun. That's fine, but fun is all you have. It's just a game. What's so important about it?

But consider that for the first time you can play when *you* want to; *whenever* you want to. You don't have to persuade someone else to play a game because you want to—or fight off someone's importunities when you don't want to.

The computer is at your service and has no will of its own. It doesn't sigh and look pained when you make a dumb move, or sneer when you lose, or make excuses when you win. Nor does it get petty and refuse to play if you lose too often—or win too often. It doesn't even sarcastically ask you if you intend to move before dying of old age if you take a few moments to think out a knotty combination.

You've never played chess under such favorable conditions before. You can take your time. You can even put a game aside and return to it later, for the computer will wait. And if the computer's program makes it no better a chess player than you are, you will win half the time.

In fact, you will catch on to some of the computer's ways of playing and you will get to be better yourself as you learn by experience. Then, when you begin to win most of the time, you can get a better program for the computer.

In short, while you're having fun and while you're playing a game, what you're *really* doing is learning how to play chess better. It is impossible to engage in any activity with an intellectual content, in an interested and concentrated manner, without learning. And if a computer makes it possible for you to engage in such activities on *your* terms, in *your* good time, and in *your* way, it becomes impossible for you to do so in anything but an interested and concentrated manner; so you learn.

The computer is the most efficient educational device ever invented because it makes it impossible for you not to learn. Teachers can be insensitive, books can be dull, but computers produce a system in which only *you* count, and you cannot be insensitive or dull to yourself.

At the present rate of computer advance, the time will soon come (always assuming our civilization does not crumble through our own follies) when any household that wishes can have a computer terminal, and can have available to it a complex and thoroughgoing system for information retrieval. This implies a number of things:

You can get what you need for daily life—weather information, the specials and prices at local stores, news and sports headlines.

You can get what you need for daily business—stock market reports, office data, letters received and sent out. It can enable you to do your work at the office or at the plant electronically, or hold conferences by closed-circuit television if your system is complex enough.

Most important, it can give you information that you just happen to want for no other reason than that you want it.

The last is the most important factor of all. All the other things a computer system can do merely make easier something we have always been able to do less conveniently. We could always make a telephone call, or buy a newspaper, or go to the office or plant.

But casual information? Curiosity information?

You might have books, but surely not every book in the world. You might go to the library, but it won't have every book either, and trying to find one that might be helpful and then working through it would be enough to make the task so difficult as to kill the curiosity.

Yet the day will surely come when the world's libraries, the world's entire store of information, will be computerized; when elaborate retrieval systems will be established so that key words can, with little delay, produce reference lists and, for that matter, the reference content itself if the request is specific enough.

If you want to know when Peter the Great was born, or what the Donation of Constantine was, or what Bessel functions might be, or what the latest information on Saturn's satellites is, or who holds the record for the total number of no-hit games pitched in a career, or how much 562 divided by 75 is . . .

Why not?

Moreover, one thing will lead to another. An answer may well give rise to further curiosity and take you off on side-issues.

Isn't this what a teacher at school is for?

No. Teachers have limited knowledge and they have thirty other students to take care of. They quickly tire of side-issue questions.

Isn't this what books are for?

No. A book can only tell you what it tells you. If something in it stirs a question within you that the book doesn't deal with, you must find another book that does, and this you may not be able to do.

In the case of a computer terminal which is connected with a global computerized library, your first innocent question may lead you to longer and longer searches for information. You may end with passages from half a dozen books which you may decide to preserve as printouts for rereading at leisure. And even then, you deal only with the significant portions of books, as the computer, prodded by your questions, refers you to this book and that to suit your needs.

To suit *your* needs.

You will be learning without even knowing you are learning because we don't call it learning when we are doing something we *want* to do, any more than we call it work. Learning is something that someone else wants you to do, according to a curriculum imposed upon you at a place, time, and speed also imposed on you. At least that is what we have been trained to think learning is.

Will computerized self-education work?

There's no way in which it can fail to work. Self-education has worked in the past for highly motivated, unbearably curious, unendingly ambitious people. Using only occasional books and incredible drive, the Michael Faradays, Thomas Edisons, and Abraham Lincolns of the world have risen to great deeds.

But where is the cosmic law that says the process must be made so difficult that only top-rank geniuses can overcome the obstacles?

Suppose everyone has a chance at any book or any piece of information just by signalling for it. People with infinitely less on the ball than the Faradays, Edisons, and Lincolns could get somewhere, do something. They would not be geniuses, but they would at least work more nearly at their top, and that might well be very good.

And how many people would want to know anything at all? Aren't most people just blanks?

Not so. People resist learning because they rarely have any chance to learn on their own terms. Youngsters in school are taught unimaginatively, and by rote, matters concerning which they are not curious; or matters in which they might be curious were it not that curiosity was never aroused; or, worst of all, matters in which they were once curious but in which that curiosity was killed.

But then, if people use computerized information to learn exactly what they want to learn and no more, who's to say that that kind of learning will be of any importance whatsoever? What if hordes of people are curious only about baseball scores, or about the private lives of movie stars?

Even so, one thing leads to another. Baseball scores may lead to an interest in how one throws a curve, which may then lead to a curiosity about the physics of moving bodies. The private lives of movie stars could lead to a serious interest in the dramatic arts.

And if it doesn't?

Then at the worst, we have lost nothing, because all the effort to teach people "worthwhile" things is wasted if the people being taught don't want to learn. Look about you! Every person you see went to school and studied mathematics, history, geography, literature, and all the time-honored subjects—and the chances are you couldn't scare up enough knowledge among all of them put together to pass a fourth-grade quiz.

Will computerized education create an ingrown culture in which everyone will hunch over computer terminals and be interested only in what *they* are interested in? Will all interhuman contacts be lost?

No. In the first place, not all the things one is curious about can be obtained from information already frozen. There are some subjects that require the outside world—laboratory work, fieldwork, public speaking, drama, sports.

Computer-teaching will therefore not utterly replace conventional teaching, nor should it. Indeed, students will welcome human interaction more because it will not be the only mode of instruction open to them. They will find the classroom more interesting knowing that anything that arises out of it to pique their curiosity might be amplified by the computer.

In the second place, even if conventional teaching did not exist, computer-teaching would not necessarily build a wall around a student fascinated by his own curiosity. That is not the way it works.

We already have a device that is capable of building a wall around a person. The television set has its devotees who will sit passively watching for hours every day. Will this prevent human interaction? It could—but not necessarily.

Few programs have so caught young people's entire imagination as "Star Trek." It has become a virtual cult—but it spawned conventions. The first of its kind was thought by its organizer to be likely to attract 250 people: it brought in 1,400. The second was geared for 2,000 and attracted 4,000—all of them excitedly interested in each other because they all lived in the same fan world.

The enthusiast is sure to be a missionary. Any youngster who, through his exploration of the world of information, finds some esoteric field that utterly fascinates him will seek out others who may be equally fascinated. Failing that, he will try to teach and convert.

That this should be so is exciting indeed. Given enough time, any student who will find he has wrung out of a field all that the computer can find will start trying to make contributions of his own. If interest is sufficiently fierce and curiosity sufficiently unbounded, research will begin.

Yet even after all of this, we haven't plumbed the deepest significance of computer-education.

Earlier in the essay I said that the advance of computer-education depended on the hope that our civilization would not crumble through our own follies.

One of the follies that would inevitably destroy us all would be that of continuing to allow the population to increase in number indefinitely. Four and a quarter billion people are now on Earth; and with declining reserves of food, water, and energy, the population is still increasing by 185,000 each day.

The world is coming to realize the danger, and the cure. It is necessary to lower the birth rate. Western Europe has about achieved zero population growth in this way. The United States is approaching it. China is fighting hard to achieve it. Even the Third World is waking to the peril.

Suppose we *do* reach the cure. If we have a low-birth-rate world-society for the first time in history, and combine it with a high technology and advanced medicine, we will also have, again for the first time in history, a quickly aging population. We will have the largest percentage of people who have reached the autumn of post-maturity, and the smallest-ever percentage of people in the spring of youth; many old and few young.

It is something that some might fear, for it is part of popular wisdom that old people are crotchety, querulous, dull, and without vision. Only the young, supposedly, are brave, strong, creative, driving, and productive. Will the world, then, having escaped destruction through the bang of overpopulation, retire to the slower and perhaps more harrowing death through the whimper of old age? Are those the only two alternatives that can possibly exist?

I think not. Our opinions of the old are the product of our system of education, which is confined to the young. What's more, it treats the young in so inefficient a way that the young are repelled by it, escape from it as soon as they can, and then never return to it because they view it as a hated childishness they have outgrown. We create millions of old people in this way whose only experience with education is remembered with scorn and bitterness. And even if there are old people who somehow would like to learn something—anything— we do not have strong social institutions to accommodate them.

But how will the computerized education that is now dawning in the world accommodate them?

If it is possible for youngsters to satisfy their curiosity by making use of the accumulated knowledge of the world through a device that will cull that knowledge and retrieve specific items on command, why should it be only youngsters who will use that device? Or even if it *is* only youngsters who do so at first, because those who are no longer young have been ruined past reprieve by conventional education, why should the young stop doing so at some fixed age?

People who enjoy golf, tennis, fishing, or sex when they are young do not willingly stop because they reach the age of thirty-five, or forty, or fifty, or any age. They continue with undiminished enthusiasm for as long as they are physically able to do so.

So it will be with learning.

It may seem strange to place learning in the class of pursuits which we associate with fun and pleasure, but learning *is* fun. For those who, even in our own inefficient educational system, find themselves enjoying it, learning is the greatest pleasure in the world, one that outlasts all the others.

How much more so would it be when education is completely under one's own control, when one can learn what one wants, when one wants, where one wants, and how one wants; when one can learn something today and another thing tomorrow at will; when one can follow the track of curiosity, at one's own speed and choice, wherever it might lead?

While a mind is exercised and always freshened with new interests, it will not age. Death comes at the end when the physical machinery of the body breaks down, and the mind will die with it, but it will die active and vigorous.

The time is coming; the home computers are with us; we will be growing more familiar with them and learning ever better how to use them; and they will be connected more and more thoroughly to the varieties of information potentially available to people.

The result?

There will be greater intellectual depth and variety to humanity than the world has ever seen. It will be an exciting world, a bubbling and effervescent world in which hosts of interests will compete with each other, and human beings will race each other to be the first with a new finding, a novel idea, a better book, a more illuminating truth, a cleverer device.

They will look back on everything that existed before the age of the home computer as a time that belonged to the infancy of the human species; and they will consider the home computer the path to adulthood for humanity.

But when? How much will we have accomplished of all this by the year 2000?

That depends on how much we will allow ourselves to accomplish, on whether we have the good sense and the will to allow our civilization to continue.

If we choose correctly, however, then what change does occur, large or small, will inevitably be (it seems to me) in the direction I've indicated.

—— 51 ——
TECHNOLOGY, YOU, YOUR FAMILY, AND THE FUTURE

The future is rushing toward us, whether we like it or not, and women must not be caught unprepared by it. It is going to affect the home and the family in very fundamental ways—there's no doubt about that—and it is also going to affect the status of women generally.

All through history, the nature of the future has been affected to some extent by technological change. Through most of history, however, such change has been initiated, and has spread, and has had its effect, so slowly that it could safely be ignored in the course of a lifetime.

The pace of technological change has, however, steadily quickened, and, to a greater and greater extent, the nature of the future has been determined by such change rather than by anything else. Technological change began, increasingly, to determine the direction in which political, economic, and social institutions changed.

By the time the Industrial Revolution arrived in the 1770s, with the invention of the steam engine, change could be seen with the naked eye, so to speak, and in the twentieth century the speed of change became a high wind.

It is now a hurricane, and its name is Computer.

We are rapidly becoming a computerized society. The computers, which for a generation we have associated with industry and the universities, are entering the home. This has come about because of the development of the "microchip," which has made it possible to cram a great deal of capacity and versatility into a box small enough to fit on a desk and cheap enough to fit into a pocketbook.

The same microchip has made it possible to build machine tools that are so intensely computerized as to be capable of doing a variety of tasks that, till now, only human beings could be relied on to do. These are the "industrial robots" which are in the process of transforming our factories right now. They don't look like the robots of fiction and have not the vaguest resemblance to the human shape. That, however, is coming, too. The "home robot," which may be increasingly humanoid in appearance, is in the process of development, and robots that can see and talk and do jobs about the house may very likely be with us in a decade or two.

How will this change the home and the family?

In a number of ways, some superficial and easily foreseen, and some, quite deep, that might produce unexpected results.

For instance, at the superficial level, you might be able to computerize routine and repetitive labors about the house: settle your accounts, balance your checkbook, keep track of your Christmas cards, better control the lighting and heating of the house or the watering of the garden. A home robot might be able to run a vacuum cleaner or a dish washer, or hang up coats.

None of this is really very vital. It saves you time when everything is running well, but if the computer or the robot is "down," you may wonder how worthwhile it all is. (As when your car is acting up and you can't get to the shopping center and you think, irritably, that it would be nice if, like your mother, you could just drop in at the corner grocery.)

Move deeper, however. . . .

The computer gives you access to information, and accelerates the transfer of that information. A home computer could offer a way of keeping you abreast of the news, weather, and traffic conditions. All this, to be sure, is done by radio and television, but the computer will do it on demand, at your convenience.

What's more, you can react to the information. If you are hooked up to your bank, you can not only find out the state of your bank balance, you can arrange to transfer funds in and out. If you are hooked up to a supermarket or a department store, you can not only study commodities and prices, you can make your purchases and arrange for delivery. If you work in an office, you can make contact with that office from your home, dictate letters, receive information, supervise procedures, and hold conferences.

In short, much of your daily work, whether home-related or job-related, can be done without stirring from your living room. This will obviously hold for your husband as well as for yourself. This will introduce subtle but far-reaching changes in your commuting habits, for instance, or in the way you must organize your time. It will also mean increasing the time with your family (multiplying opportunities for closeness—and for friction, too).

There are those who argue that this "work-from-home" will never be practical; that human beings are social animals who want to interact with other human beings, who want to get away from the narrow limitations of the home and gather in stores and offices. Even if that were so, no one suggests all-one or all-the-other. On pleasant days, when you want the feel of humanity about, you can go to the store or office, but on days when you are tired, or the weather is bad, or you simply don't want to bother, you can use the computer. The computer doesn't *compel* you to do anything, but it does offer you a choice.

Still deeper. . . . The computer is an educative device. It can be hooked up to libraries of information of all sorts and can, in effect, answer questions or refer you to books or documents or any other reference material. It would, on demand, produce any of these, page by page, on the screen, and print out those pages you would wish to keep for a period of time.

The process of learning is a delight when it is not forced. When you are dealing with something you are curious about, and want to know—when you can study it at a time and place and pace of your own choosing—and when you can skip freely from subject to subject as your fancy chooses, you will be surprised at the pleasure of it.

This will be important in connection with your children. We know the fascination of computer games already, and of how addictive they can be for youngsters, but individual games are, of their nature, limited, and lose their charm eventually. There is only so much that can be sucked out of them. Learning is a kind of game, too—but its charm is infinite in scope.

Computers will not replace schools; the need for interpersonal contact in many fields will remain; but computers will certainly be an all-important supplement that will change the face of education forever.

The important thing to remember is that it will be not only your children who will have the entire corpus of human knowledge open to their browsing at will, but *you* will have it as well.

In human history, education has largely been reserved for the young, and it is customary to feel at a certain, not-very-old age that one's schooling is finished. With a computer forever at one's disposal, however, and through it, all the world, there is no reason to put an end to one's pleasure at any time. Learning will indefinitely continue in those directions that delight you, and the human brain, kept in condition through the decades, will remain lively and creative until extreme age withers it physically.

With this in mind, we come to the deepest use for computers, that of finally raising the role of women in the world to full equality with that of men.

We have reached the point where we understand that in an ideal world there should be no such thing as "women's work" or "men's work"; that each individual, regardless of sex, should be allowed to work at whatever he or she can do well, and for a return to be determined by the nature and quality of the work done and not by the sex of the worker.

In response to this present attitude, women are flocking into positions that very recently were reserved for men. The legal profession is an example. Those fields which are becoming feminized, however, have one thing in common. They do not involve mathematics or the physical sciences.

For a variety of social and psychological reasons (not excluding the still-continuing masculine hostility in these fields) women are not attracted to mathematics and physical science. Men are only too willing to believe that women have an inherent lack of aptitude for these subjects, and unfortunately, many women, in flinching away, console themselves with the same myth.

In a world in which mathematics and science are very important, however, leaving those fields in the hands of men only seriously limits the extent to which women can achieve the goal of equality. As it happens, the computerization of the world is going to multiply those jobs that involve the manufacture, maintenance, and repair of computers, the programming and reprogramming of computers, the design of new computers, research into artificial intelligence and the neurochemistry and neurophysics of the human brain. All of these will involve much in the way of mathematics and science. If women, because of this, abandon the world of computers to men—they abandon the world to men.

But it is computers, too, that offer the solution. If girls at an early age are not directed away from science and mathematics as being somehow not "suitable" for girls; if, rather, they are encouraged; if, with the help of teaching-computers, they can study by themselves in their own way, without the pressures of outmoded superstitions forcing them back; many will find that these subjects are not particularly difficult. By the time they are adult, it will no longer be possible to stop them.

And once science and mathematics are "sex-blind," then the whole spectrum of achievement will be sex-blind, and humanity for the first time in its history will be able to use *all* its talents, rather than only half.

——— 52 ———
SHOULD WE FEAR
THE FUTURE?

It is always possible to fear the future, for who knows what dire events might not lie ahead?

It is always possible to have hope, however, for who knows what solutions the inventive minds and keen insights of human beings might not produce?

At the present moment, what hope there is must rest on the computer revolution. Surely computerized machines (robots) would increase production and make it more efficient. Computers themselves, growing steadily more versatile and capable, would help administer human economy and human society more efficiently and justly. It would surely help us solve problems that at present stump us, so that those social disorders that help create unrest, alienation, and violence might at least be ameliorated, and society made healthier as a result.

But what if the very means by which we expect to have our problems solved will themselves create problems that could be worse than any that already exist? What then?

In other words, might the computerized society we now face be a source of agonizing fear rather than of hope?

There are some, certainly, who fear such a society, and with reason. A robot on an assembly line might do a particular job more efficiently, more tirelessly, more satisfactorily, and far more cheaply (once the capital investment is made) than three human beings together could possibly do, and that is good; but the three human beings would then be out of work, and that is bad.

Industries, in periods of recession, might lay off employees by the tens of thousands and seize the opportunity to computerize. In periods of economic recovery, computerization would then allow expansion of production without the need of rehiring employees. The result might be a permanent, and growing, unemployment problem in the nation and the industrial world that might have no solution, and that, in the end, would destroy the stability of society.

Some might dismiss this danger.

They might argue that history makes it quite plain that advancing technology is, in the long run, a creator of jobs and not a destroyer of them. The coming of the automobile put a number of blacksmiths and buggy manufacturers out of business, but created a far greater number of automobile-related jobs. In a more general way, the Industrial Revolution, despite the miseries of the early decades of the new factory system (low wages, long hours, bestial working conditions, and the heartless exploitation of women and children), eventually brought a relatively high standard of living to countless millions that would never have experienced it otherwise.

In answer to this, there is the point that the jobs that are created are usually quite different from those that are destroyed, and it is not always possible for a person who has lost an old-style job simply to take a new-style one. There is the problem of retraining and reeducation, which takes time and money—and therefore compassion. It would require a willingness on the part of those fortunate ones who are not dislocated to give up some of what they have (in the form of taxes, probably) so that programs for the relief of the unfortunate might be financed.

To be sure, governments and societies did not undertake this responsibility in the early period of the Industrial Revolution, and yet the world survived—or, at least, so it might be argued.

But did it survive without serious harm?

Even if we dismiss the incredible horrors of the early decades of the factory system, we will have to admit that it was the callousness of employers and the governments they controlled, their indifference to misery, that led to the growth of Marxist doctrines that so many Americans now fear and oppose. (It took a long time for many employers to discover what Henry Ford was to demonstrate—that *better* working conditions were the route to company success).

Furthermore, the situation today is different from what it was two centuries ago in important ways. For one thing, the rate of technological change is far more rapid now than it was then, so that the social and economic dislocations involved in the change will become much more intense in a far shorter time than it ever did in the earlier period. Secondly, in the industrial nations generally, those in the lower regions of the economic pyramid have grown accustomed to a government philosophy that includes compassionate consideration for the unfortunate and will not be ready to submit to indifference.

Yet all this does not necessarily mean we must look forward to certain disaster.

I think that society in general has learned a great deal in the two centuries that stretch between the invention of the steam engine and the collapsing economies of the Great Depression. We have learned that governments cannot turn their back on the miserable and long endure.

Reagan in the United States and Thatcher in Great Britain have indeed in recent years attempted to reverse the trends of a half-century in order to comfort the rich, while leaving the poor to comfort themselves. That this has not brought instant reprisal is due to the fact that the cushions instituted by earlier liberal governments remain in place and are, in fact, irremovable. As an example, Reagan's foolish effort to weaken social security brought instant rejection by a rare unanimous vote on the part of senators terrified at the prospect of the loss of their own reelection social security. Thatcher survives high unemployment because the unemployed can live on various benefits which it would be instant suicide for her to withdraw.

Nor is the conservative policy likely to last long even so. That Thatcher was recently reelected with less than half the vote (hailed as a landslide by a press that couldn't count) was entirely due to the jingoism induced by an unremarkable Falkland victory over a nation with a strictly amateur military force, together with the nearly even split of the opposition. And if Reagan is reelected in 1984, it will represent the ability of a Hollywood smile to obscure brainlessness.

I am certain, therefore, that as the pains of the transition from a noncomputerized past to a computerized future become more intense the western world will take the necessary measures to care for those who will need care. The philosophy of compassion will return, for it will be the clear path of long-range self-interest. There will be jobs programs and government-sponsored retraining measures.

Most important of all, the government will surely recognize that nothing is more crucial than the education of the young.

This is not something new. In the preindustrial past, an agricultural economy did not require literacy. Peasant labor on primitive farms could be carried on with no education greater than that which could be passed on from parents to children by imitation. An industrialized society, however, requires more elaborate skills; and to pass these greater skills onward across the generations requires more than such imitation. It is no coincidence, therefore, that government-sponsored free education for the general population became public policy in those nations which aspired to industrialization. They could not have maintained the industrialization otherwise.

A "postindustrial" computerized nation will once again require a higher level of education. Fortunately, the fact of computerization itself will make such a thing possible. Computerization increases the ability to transmit and handle information. When computer outlets are available in the average home, when libraries are computerized and within reach, when questions can be answered and instruction can be offered by carefully programmed computers on an individual basis, youngsters (and, indeed, people of all ages), finding themselves able to learn at their own speed and in their own way, will devour learning and actually *enjoy* education. The disadvantage of a frozen curriculum that takes no account of individual differences will disappear.

If the transitional period between a noncomputerized and a computerized society is frighteningly rapid and painfully intense, it will also be brief—one generation long. The expensiveness of compassion for the displaced, while hard to bear by those unused to loving-kindness, will at least not endure unbearably long.

And the new society will have its pleasant surprises. The coming of the public school showed that literacy was not the property of a small intellectual elite after all; they showed that *nearly everyone* could learn to read and write. In the same way, the coming of computerized education will show that creativity is not the property of a small intellectual elite after all; they will show that *nearly everyone* has areas of creativity.

Nor will this newly creative humanity be forced (as in the past) to blunt that creativity through the necessity of doing dull and repetitive labor merely in order to earn a living, labor that will stultify and reduce the mind to subhuman stupidity. Not a chance, for it is precisely the dull work, the repetitive work, the nondemanding work, and the mind-destroying work, for which computers and robots are designed, and from which human beings will therefore be insulated.

Those who think of the computer as a dehumanizing influence are totally wrong, completely wrong, 180 degrees wrong. Humanity has lived cheek and jowl with dehumanization for thousands of years, and it is from this that it will be freed by the computer.

But is it reasonable to think in this positive fashion about the future? Is it perhaps unrealistic to accept the optimistic scenario I am projecting—a transition period which, while intensely painful, can be cushioned and will be brief, and which will be followed by a radically new, and much better, computerized society?

After all, is it not possible that the ills which now infest society—a growing overpopulation, a general dwindling of resources, an overpowering miasma of chemical and radiational pollution, and the everpresent threat of terrorism and nuclear war—will destroy us all before the glowing future, or *any* future, can be realized?

Disaster *is* possible. Society may collapse because of human hate and suspicion, fostered by the stupidity and blindness of the world's leaders. Still, if such disaster is *possible,* this does not mean it is *inevitable* and, indeed, I see signs that it may be avoided.

In the nearly four decades since the United States used two atomic bombs against an already-defeated adversary, no nuclear weapons have been fired in anger. The United States resisted the temptation to use them in Vietnam and the Soviet Union is resisting the temptation to use them in Afghanistan. World opinion against nuclear weapons is steadily intensifying, and will eventually become insuperable.

Overpopulation is recognized more clearly each year as a deadly danger, and, increasingly, efforts are being made the world over to limit the birth rate. Population is still increasing, but the rate of increase has been dropping over the last decade.

In the same way, international concern over the other major dangers has been rising. Reagan's attempt to fight "radical environmentalists" resulted in humiliating failure, and a new appointee had to admit that the ex-actor had "misread his mandate."

In addition, we are moving into space, thanks, to a large extent, to the abilities of advanced computers. The formation of space settlements and space industries is coming nearer to realization with each passing year. In space there will be new sources of both energy and materials, sources that, with proper handling, will be inexhaustible. There will be new space for humanity and new opportunities for expansion.

All the evils that point to the possibility of the destruction of civilization can be fought and, to some extent, *are* being fought; and if they can be held off long enough, they can be ameliorated to the point where they will no longer be threatening.

We will not only face a computerized future, but an expanded more-than-Earth future.

And yet one danger remains, a danger intimately related to the very nature of computers. Is it possible that computers, forever advancing, may become so capable, so versatile, so advanced, that they will reach and surpass human abilities? Will they make human beings obsolete? Will they take our place and reduce all optimism to nothing, producing a world that will consist only of computers and in which human beings will find no place?

No, I think there is ample reason for believing this will never be so, but it will take another essay to explain my reasons for believing that.

SHOULD WE FEAR
THE COMPUTER?

It is possible to welcome the computer (as I did in the previous essay) as the means by which humanity will be liberated of all drudgery, mental as well as physical, so that people will be free to become utterly human and creative. But might not the computer go too far? Might it not develop so many abilities, prove so versatile and (eventually) intelligent, that it will supersede and replace the human being entirely? In short, will human beings become obsolete?

If one wishes to be cynical (or, perhaps, simply rational) the proper answer might be "Why not?" If computers become more capable than human beings, if robots can be manufactured with stronger, better bodies and a more intelligent computerized brain, why shouldn't they replace us for the same reasons that mammals replaced reptiles as the dominant form of vertebrate land life on the planet, or for the same reason that *Homo sapiens* has come to dominate Earth as no single species has ever done before?

One might even argue that all of evolution has proceeded as it has, by trial and error, and by the slow and inefficient mechanism of natural selection, only so that a species might finally evolve that would be intelligent enough to *direct* evolution and create its own successor, which would then continue the chain so that change would proceed in centuries that would earlier have taken eons—toward some goal we cannot guess.

In fact, if we indulge ourselves in unrestrained misanthropic musings, we might even come to the conclusion that it is not the coming of the displacing computer that we must fear, but the fact that it might not come quickly enough. The human record is a dismal one. Our treatment of one another, of other species, and of the very planet we live on, has been brutal in the extreme. We now have the power to destroy civilization and even, perhaps, the ability of Earth to support complex life-forms; and we can have no confidence, on the basis of past history, that we will refrain from such destructive behavior. We *should*, perhaps, be replaced by more intelligent thinking machines which might very well prove more rational and humane.

The question, however, is not whether computers (and their mobile versions, the robots) should or should not overtake and replace us, but whether they *will* or *will not* do so.

It is not at all easy to say that something will or will not happen (far less easy than to argue moral imperatives), but I will take a chance and suggest that it will *not* happen. Computers will *not* overtake us. We may well destroy ourselves, but computers will *not* do that job for us.

Why do I say that?

Well, for one thing, the human brain is not that easy to match, let alone surpass. The brain contains 10,000,000,000 neurons, or nerve cells. Each of these is connected to anywhere from 100 to 100,000 others. What's more, each neuron is not merely an on-off switch (which would in itself be enough to make the situation enormously complicated) but is an ultracomplicated physical-chemical system that we are not even on the brink of understanding. The total complication is almost beyond expression, let alone beyond comprehension.

An important difference between a computer and a human brain, then, is that the computer is completely defined. One knows exactly what a computer can do and just how it can do it. On the other hand, one does not know precisely what a human brain can do, and one certainly does *not* know how it does it.

Thus, we know how a computer memory works, we know exactly what its contents are, and how it retrieves a particular item among those contents. We do not know how the memory of the human brain works, nor what its contents may be, nor how it retrieves a particular item.

With respect to the human brain, in fact, the whole operation of memory and recall is an enormous puzzle. For instance, my own memory and recall is (and always has been) excellent, and I have amused myself, now and then, by trying to sense how it works. Naturally, I have failed.

For instance, I know that "claim" is an English word and "clain" isn't; that "career" is an English word and "creer" isn't. I can look at uncounted numbers of letter-combinations and say, with confidence, that they are not English words and I will very rarely be wrong. But how do I know that? Do I somehow leaf through all the English words in my memory and note that a particular letter-combination is not included among them. Is it possible for my brain to work that fast? If not, is there an alternative? I don't know.

The situation grows even more mysterious if I ask myself how I write. I write, for publication, at least 500,000 words a year and do that well enough to get them published, too. Clearly, the words are only publishable if chosen with a certain degree of aptness and then put together with a certain degree of elegance. How do I do that? I don't know.

I write, on the whole, as quickly as I can, choosing each word as I need it with almost no hesitation, and they end up being (more or less) the correct words in the correct order.

I am anything but unique in this respect. There are other writers, some who are better than I am. There are other creative artists in many fields. There are people who work with their hands and put things together. There are athletes who can perform difficult feats with marvelous neuromuscular coordinations. In every case, they do things, almost without thought, that are incredibly difficult, judging from the fact that we cannot describe just how they are done.

I have never ceased to marvel at a baseball outfielder, for instance, who, at the crack of a bat, turns and races to a particular place, then lifts his gloved hand and plucks the ball out of the air. It isn't the fact that I don't see how the devil he can so quickly and unerringly judge the path of the ball that astonishes me. The marvel is that *he* doesn't see how the devil he does it.

Even people who are quite ordinary, who are not what the world would call "creative," who are not even bright, do a large number of things every day— without any conscious effort—and do it by some means that neither they, nor anyone else, can explain.

How then can we program a computer to duplicate those processes of the human brain that we do not understand, even very primitive processes that no one thinks of as remarkable (because we take our brains so for granted).

Think of it. You can look at the letter A in its capital form, or as a small letter, or in italic, boldface, or any of a hundred styles of print; or as a handwritten capital or small letter in the separate handwritings of a thousand different people; and in each case you would recognize it, in a very short time, as the letter A. You would probably be surprised if anyone told you that you were doing something remarkable—but we can't get a computer to do it.

You can recognize any of an infinite number of voices, and do so at once even when it is distorted by the imperfections of a telephone or a recording device, and even, sometimes, when you have not heard it for a long period of time. Why not? The ability to do so doesn't astonish you—but we can't get a computer to do it.

What we *can* get a computer to do is, essentially, simple arithmetic. Any problem, however seemingly complex, which we can somehow break down into a well-defined series of simple arithmetical operations, we can get a computer to do. That the computer can amaze us with its capabilities arises out of the fact that it can perform these operations in billionths of a second, and can do so without any chance of making a mistake. It can't do anything we can't do, but it would take us billions of times longer to do it, and we would almost certainly make many mistakes in the process.

Surely, the reason that we do computer-stuff so poorly is precisely because it is unimportant to us. The human brain is not designed to plod away at infantile calculations. These have only become important, superficially, as civilization has introduced taxation, commerce, business, and science; and even so they remain unimportant in the context of what anyone would consider the major concerns of living. The business of the human brain has always been, and still is, judgment and creative thought, the ability to come to a conclusion on the basis of insufficient evidence, the ability to think philosophically, insightfully, fancifully, imaginatively, the ability to extract beauty, excitement, and delight out of the world that surrounds us, and out of what we ourselves create.

Can we get computers to do what the human brain does, even though we can't reduce our human skills to simple arithmetical operations? Can we perhaps learn to make a (perhaps simplified) map of the human brain and build a computer that mimics that map as a Tinker-toy construct might mimic a skyscraper?

Could we perhaps not so much build a computer that can do what the human brain does, but build a much simpler computer that can at least show the capacity to *learn* as a human brain does? Could it, after a time, learn enough to instruct human beings on the design of a computer closer to the human brain, that could, in its turn, learn enough to help build a computer closer still, and so on?

In short, might we not build a computer that works according to principles that we might possibly not understand in detail, because it is based in part on what earlier computers have learned rather than what we have learned ourselves? Despite that, would it then do the job just as human beings do?

Even if this were possible, the chances are that we would not want to do it—not because we were afraid to do so, but, more likely, because it would take too much effort for any good it might do us.

Consider the analogous situation of machines that move. For a long time we have had inanimate devices that could move on land. Until the early nineteenth century, they moved only if they were pushed or pulled by human beings or animals, but in recent times they could "move by themselves," thanks to the steam engine, and then to the internal combustion engine.

These devices do not move, however, as human beings move. They made use of the wheel and axle and rolled; they did not walk. Rolling is useful because it does not involve periodic lifting; the device and any objects it carries move along at a steady level. In walking, there is a lifting process at every step. On the other hand, rolling requires a continuous, more-or-less flat and unobstructed way, so that roads must be built. And for modern devices, those roads must be paved and are difficult to maintain. In walking, one steps over minor obstructions, of course, and roads need be little more than paths. Why not, then, build machines that walk rather than roll, that are human (in that respect) rather than nonhuman?

Obviously, it is not done because a mechanical walker is so much more difficult to build than a mechanical roller, and because even when built (and some have been), the mechanical walker is too clumsy to be much good for anything but the most specialized uses.

In the same way, even if a human-brain computer were built, it is very likely that it would take enormous effort, enormous time, and end up as a very pale substitute for the human brains we produce by the billions in the ordinary way.

Might this not be merely unadventurous thinking? If a computer is designed to design one better than itself which can then design one still better, and so on, might the process not go out of control? Could we build not only a human computer, but a superhuman computer too powerful for us to stop it? Might we not find ourselves slaves, or worse, before we even realize we are in danger?

This makes a very dramatic scenario for science fiction (I've even used it myself, except that my advanced and out-of-control computers are always benevolent—angels rather than demons) but I don't think it would happen in real life. The human brain is so advanced in its way, and progress toward it would have to be so tentative, that I think that human beings would recognize the danger long before it arrived and would "pull the plug." It is likely, in fact, that humans are already so suspicious of computers that they will see danger where none exists and pull the plug unnecessarily, out of fear that they will be caught napping.

So I think it is extremely likely that computers and human beings will continue along parallel paths, with computers continuing to do what they do so much better than we, while we continue to do what we do so much better than they. There will always be room for both of us, so we will be symbiotic allies rather than competitive foes.

What we really need fear is not that through fatuous overconfidence we will nurture a master and supplanter, but that through foolish suspicion we will fail to avail ourselves of an absolutely necessary helper and friend.

—— 54 ——

WORK CHANGES ITS MEANING

Periodically in history there come great transition periods in which work changes its meaning. There was a time, perhaps ten thousand years ago, when human beings stopped feeding themselves by hunting game and gathering plants, and increasingly turned to agriculture. In a way, that represented the invention of "work," the hard daily labor designed to insure food and the wherewithal of life generally.

"In the sweat of thy face shalt thou eat bread," says God to Adam as the first man is expelled from the garden, and the transition from food-gathering to agriculture is symbolized.

Then in the latter decades of the eighteenth century, as the Industrial Revolution began in Great Britain, there was another transition in which the symbol of work was no longer the hoe and the plow, but the mill and the assembly line. The aristocrat no longer cultivated a pale skin as his mark that he did not have to go out into the fields to labor; he cultivated a tan to show that he did not have to stay indoors to labor.

And now we stand at the brink of a change that will be the greatest of all, for "work" in its old sense will disappear altogether. To most people, "work" has always been an effortful exercising of mind or body, bitterly compelled to earn the necessities of life plus an all-too-occasional period of leisure in which to rest or "have fun."

People have always desperately tried to foist off work on others: on human slaves, serfs or peasants, on hired hands, on animals, on ingenious machines. With the Industrial Revolution, machinery powered first by steam, then by electricity and internal combustion engines, took over the hard physical tasks and relieved the strain on human and animal muscles.

There remained, however, the "easier" labor that did not require muscle, and that machines, however ingenious, could not do—the labor that required the human eyes, ears, judgment, and mind. If this work did not require huge effort, bulging muscles, and sweat, it nevertheless had its miseries, for it tended to be dull and repetitive. Whether one works at a sewing machine, an assembly line, or a typewriter, day after day, there is always the sour sense of endlessly doing something unpleasant under compulsion, something that stultifies one's mind and waste's one's life.

Although such jobs, characteristic of the human condition in the first three-quarters of the twentieth century, make too little demand on the human mind and spirit to keep them fresh and alive, they have made too much demand for any machine to serve the purpose—until now.

The electronic computer, invented in the 1940s and improved at breakneck speed, was a machine that, for the first time, seemed capable of doing work that had till then been the preserve of the human mind. With the coming of the microchip in the 1970s, computers became compact enough, versatile enough, and (most important of all) cheap enough to serve as the "brains" of affordable machines that could take their place on the assembly line and in the office.

This means that the dull, the repetitious, and the mind-stultifying will begin to disappear from the job market—is *already* beginning to disappear. This, of course, will introduce two vital sets of problems—is *already* introducing them.

First, what will happen to the human beings who have been working at these disappearing jobs?

Second, where will we get the human beings to do the new jobs that will appear—jobs that are demanding, interesting, and mind-exercising, but that require a high-tech level of thought and education?

Clearly, there will be a painful period of transition, one that is starting already, and one that will be in full swing as the twenty-first century begins.

The first problem, that of technological unemployment, will be temporary, for it will arise out of the fact that there is now a generation of employees who have not been educated to fit the computer age. However (in advanced nations, at least), they will be the last generation to be so lacking, so that with them this problem will disappear, or at least diminish to the point of noncrisis proportions.

While the problem exists, the unemployed must, first of all, be reeducated or retrained to the best of their capacity, in order that they may be fitted for new jobs of the kind that the computer age will make available. (It is, after all, reasonable to suppose that computers will create far more jobs than they destroy—at least technological advance has always done so in the past. The only difficulty is that the created jobs are widely different from the destroyed ones.)

Consequently, as the twenty-first century opens, one of the most important types of jobs available (for those who can fill them) will be that of teaching the use of computers and all the myriads of skills involved in the design, construction, and maintenance of computers and their mobile offspring, the robots. (And of course this will, for a while, further exacerbate the second problem, that of finding all the people necessary to take part in the computer revolution).

We are sure to find among the technologically unemployed those who, because of age, temperament, or the mental damage done them by the kind of lives they have led or work they have done, are not able to profit from retraining. For these, it will be necessary to create jobs they can do, or, in the extreme, to support them—for they are human beings. And this, too, will require much work in the way of administration and thoughtful and creative humanitarianism.

But the second problem—that of finding a large enough number of high-tech minds to run a high-tech world—will prove to be no problem at all, once we adjust our thinking.

In the first place, the computer age will introduce a total revolution in our notions of education, and is beginning to do so now.

So far, in history, there have been only two fundamental ways of educating the young: individual tutoring and mass education.

Individual tutoring has always been confined to a small majority who can afford a family tutor, and even if everyone could afford one, there would not be enough qualified tutors on the basis of one to a student. Therefore, when it turned out that an industrial society needed general literacy (which a strictly agricultural society does not), there was a move toward mass education.

There are enough teachers (at least of a sort) if each one is saddled with thirty to fifty students, but the disadvantage is that students must then learn according to a fixed curriculum (which takes very little account of individual differences). As a result, every youngster learns to read and write after a fashion, but almost every one of them finds himself bored, or confused, or alienated, depending on whether the teacher is going too slow, too fast, or too "sideways," into uninteresting topics.

The coming of the computer means a return to individual tutoring on a new basis. If every home has a computer outlet and if, through these, youngsters (and oldsters, too) can tune into a thoroughly computerized library, it will be possible to probe into any field of interest about which one is curious. Schools will not vanish. They will serve to introduce subjects to their students, and they will deal particularly with those fields of study that require human interaction, but each student, having grown interested in something (and we, each of us, grow interested in *something*), can follow it up at home in an individual way and at an individual time and speed.

This will make learning fun, and a successfully stimulated mind will learn quickly. It will undoubtedly turn out that the "average" child is much more intelligent and creative than we generally suppose. There was a time, after all, when the ability to read and write was confined to a very small group of "scholars," and almost all of them would have scoffed at the notion that just about *anyone* could learn the intricacies of literacy. Yet with mass education general literacy came to be a fact. Right now, creativity seems to be confined to a very few, and it is easy to suppose that that is the way it must be. However, with the proper availability of computerized education, humanity will surprise the elite few once again.

And of course that means there will be enormous opportunity for computer engineers to organize knowledge properly, to work out teaching programs, to keep them up to date, and to revise them for greater efficiency and interest. They will be the teachers of the future, and will be more important than ever to all of us.

Granted, now, that the problems of unemployment and education will be on the way toward solution, what kind of work will there be aside from what is involved in that solution?

For one thing, much of the human effort that is today put into "running the world" will be unnecessary. With computers, robots, and automation, a great deal of the daily grind will appear to be running itself. This is nothing startling. It is a trend that has been rapidly on its way ever since World War II. As an example, the telephone industry is now so automated that when hundreds of thousands of employees go out on strike, those who use telephones are hardly aware of any change. The telephone system runs itself.

The result of the continuing trend will be that more and more working people will have more and more leisure. That is also not startling, for we have been witnessing the steady increase in leisure for a long time. This means that more and more "work" will involve the filling of this leisure time. Show business and sports will grow steadily more important. (A comment I once made, which has been frequently quoted, is: "One third of the world will labor to amuse the other two thirds.")

Hobbies of all sorts will grow steadily more important and so will those industries catering to hobbies. This is not to be deplored; it will be another way of unleashing creativity. The devil may always find mischief for idle hands to do, but the true hobbyist is never idle—in fact no one works as assiduously and with such satisfaction as the hobbyist—and so he will never be in mischief.

And the kind of work that is more easily recognized as work in the traditional sense?

There will still be much that is peculiarly human and will not be computerized except at the fringes. (I use a word-processor, which makes the mechanics of my work a bit easier, but the thing steadfastly refuses to write my essays for me. I still have to sit here, thinking.) In other words, those who are involved in literature, music, the arts, will be busier than ever in the leisure society, since their audience will grow steadily greater.

There will also be enormous changes. In business, the accent will be on decision-making, on administration. Offices and factories will be "black boxes" in which the routine details will run themselves, but in which men and women will handle the controls.

Nor will this have to be done on the spot. We will be living in an "information society" in which we will not have to transfer mass in order to transfer information. We won't have to send a human being from A to B in order that his mouth may give directions or his ear may receive directions. We won't even have to send letters back and forth in order to accomplish that. The information can be transferred at the speed of light and—thanks to computers that do the remembering, the selecting, and the transmitting—more efficiently than ever.

This means that the necessity of commuting will be increasingly a thing of the past. Business travel (as opposed to pleasure travel) will go way down. Oh, it will exist, since personal contact for psychological reasons will still be needed in business, but the inconvenient trip out of business necessity will be a thing of the past.

With computerization knitting the world into a tight unit, and with the work of directing operations and gathering information easily done from anywhere, the world will begin to decentralize. There will be an increasing tendency for cities to dwindle (this is already a recognized trend in the United States at least).

It may sound as though the twenty-first century, with its increasing leisure, its black-box offices and factories, its emphasis on long-distance work from the home, may become the kind of society that will be too secure, too easy.

Not so!

We will also be entering the "space age" with a vengeance, and perhaps the most important type of work that will be facing us as the new century opens will be that which will be involved in gaining command of the resources of space, in the transfer of industry into orbit, in the design and construction of a vast variety of space-structures, and in the making of new worlds for humanity off the surface of the Earth.

There will be incredible excitement and precious little security in that. In fact, the twenty-first century, for all its advancement, will be one of the great pioneering periods of human history.

NUCLEAR DREAMS AND NIGHTMARES

Nuclear power has suffered through the manner in which it has been perceived. To some it was an unreasonable nightmare, and to others it was an unreasonable Utopian dream.

When I first began to write and sell science fiction in the late thirties, nuclear power on a large scale was already a common idea among our group. Even though it did not yet exist in reality, the dreams and nightmares about it did. The destructive potentialities of nuclear power were thoroughly appreciated—to the full, and beyond.

I remember a story of the 1930s in which a nuclear reaction spread to the soil itself, so that a nuclear volcano, so to speak, was set off. It fizzed worse and worse with the passing years, as the nuclear fire spread, until it was plain that it was heading for an inevitable planetary explosion that would create a new asteroid belt. And indeed, when the first experimental nuclear bomb was exploded at Alamogordo, some of the assembled scientists wondered if there weren't a small chance that the atmosphere itself might be set off in a far vaster nuclear blaze than was being counted upon so that the entire planet might be destroyed.

Even the peaceful use of nuclear energy was seen as a source of intense danger. In the September 1940 issue of *Astounding Science Fiction* there appeared a story by Robert Heinlein entitled "Blowups Happen." The title tells you what you need to know. The story deals with an imaginary situation in which there was an ever-present possibility of a nuclear power station blowing up like a vast nuclear bomb.

On the other hand, after the nuclear blasts at Hiroshima and Nagasaki brought World War II to its end, another vision arose. Many people drew a happy picture of eternal prosperity through nuclear power—after it had been tamed to serve as humanity's servant. There was so much more energy in a pound of uranium than in a pound of coal or oil that visions of a never-ending energy source, of energy so cheap it wouldn't be worthwhile to meter it, dazzled us.

As it turned out, neither nightmare nor dream was real. Both were exaggerated. Nuclear bombs are incredibly horrifying, but at least the planetary body itself is safe. Far more powerful bombs than the Alamogordo firecracker have been exploded, and soil, water, and air have *not* caught nuclear fire. Nor, we now know, can they.

Again, nuclear power stations *cannot* explode like bombs. A nuclear bomb goes off only because the nuclear fire is contained for a tiny fraction of a second inside a thick shell that allows the fire's pressure to build up to the explosion point. Nuclear power stations are not so contained, and though runaway nuclear reactions might have disastrous consequences, the stations can never explode.

On the other hand, the gentle dreams failed, too. If extracting energy from uranium were as simple as throwing a lump of coal on the fire, then yes, the energy would be copious and extremely cheap. However, uranium must be purified and frequently repurified after use, a very complicated station must be built, intricate devices must be included to keep the nuclear reaction under perfect control, and radioactive ash must be disposed of. In the end, the whole thing grows expensive, so the dream evaporated into nothingness.

But one thing no one expected, a phenomenon so strange that it should be of profound interest to psychologists. There has grown up a fear of nuclear power so intense that it has brought the further development of the technology to a virtual end in the United States.

The fear was growing well before the Three Mile Island incident, but there is no question that the incident brought matters to a head.

Yet should one really wonder at the fear? Isn't it, after all, normal to be afraid of an energy source so dangerous?

But we might make the point that the danger exists, so far, only in potentiality. The Three Mile Island incident was the worst nuclear accident in the history of peaceful nuclear power, but not one person died or was even hurt as a result. In fact, there have been no civilian deaths resulting from nuclear technology in all its history.

Strangely enough, *real* deaths don't frighten us in other areas. Coal mining kills hundreds each year—but most of us don't worry about it. Perhaps we reason that, after all, only coal-miners die, and they are paid to take the risk.

Take worse cases. Automobiles kill tens of thousands of us each year, and by no means only those who get into cars and know the risks. Innocent pedestrians are killed in droves. And tobacco kills hundreds of thousands each year, including, it is thought, nonsmokers who, try as they might, cannot avoid breathing the smoke.

None of this arouses the passionate fears that nuclear technology does. In fact, any suggestions that people use seat belts or airbags, or that nonsmokers be given smoke-free areas in which to work, are viewed by many as intolerable "Big Brother" interferences with their right to commit suicide and murder.

In fact, even within the realm of the nucleus there seems to be a picking and choosing of fear. Nuclear bombs are being stockpiled in vast numbers by the two superpowers. If they are used, it will be for the sole purpose of incredible destruction. Undoubtedly, many Americans fear this, and yet the much milder danger of the nuclear power plants arouses much greater passion.

There is the problem of the disposal of nuclear wastes, to be sure, but it may not be as bad as it sounds. The intensely radioactive wastes burn off quickly. They must if they're intense. Those wastes which retain radioactivity for a long, long time are only weakly radioactive. That's why they last.

And if we kill nuclear power? Shall we then burn oil and coal instead? But oil is temporary. Yes, there's an "oil glut" now, largely as a result of conservation measures brought on by the shortages of the 1970s, but the glut won't last, I promise you.

As for coal? It will last longer, but it pollutes the air ceaselessly and shortens the lives of all of us, particularly the old and those with respiratory problems. It produces acid rain, which kills forests and renders lakes lifeless. It pours carbon dioxide into the atmosphere, and the slowly rising level of that gas is expected to produce drastic and unfavorable changes in our climate.

Yet all the fears, all the shouting, and all the demonstrations seem to be against nuclear energy—which seems almost benign in comparison.

Why?—I don't know.

Afterword: This was written before the Chernobyl disaster, but that has not changed my mind (see Essay #13). It is not that I don't think that nuclear fission has its dangers. It is that I think that fossil fuels have their dangers, too, and, in the long run, worse ones.

── 56 ──

THE NEW TOOLS IN SPACE

The first, and so far the only, basic tool for human spaceflight is the rocket. Other conceivable methods for attaining such spaceflight are either theoretically impossible (for example, antigravity); theoretically possible, but totally impractical (for example, being shot out of a cannon); or simply beyond the present state of the art (for example, making use of the solar wind as a propulsive force).

The steps by which rocketry has developed may be listed briefly as follows:

Eleventh century—Gunpowder invented in China, used for fireworks and primitive rockets.

1650—The French science fiction writer, Cyrano de Bergerac, in his romance *Voyage to the Moon* first speculates on the use of rockets for spaceflight.

1687—The English scientist Isaac Newton establishes the law of action and reaction as the theoretical basis for the rocket's ability to work in a vacuum.

1903—The Russian physicist Konstantin E. Tsiolkovsky works out the first detailed mathematical consideration of rocket-powered spaceflight.

1926—The American physicist Robert H. Goddard launches the first liquid-fueled rocket of the type eventually used to achieve spaceflight.

1944—The German rocket engineer Wernher von Braun supervises the building of the first large-scale liquid-fueled rockets capable of penetrating to the upper reaches of the atmosphere.

1957, October 4—The Soviets place the first object into orbit, marking the opening of the "Space Age."

1961, April 12—The Soviet cosmonaut Yuri A. Gagarin is the first human being launched into orbit.

1969, July 20—The American astronaut Neil A. Armstrong is the first human being to set foot on the Moon.

1981, April 12—The American space shuttle *Columbia,* the first reusable rocket ship, is launched.

This brings us to the 1980s, and we may well wonder what might lie ahead of us to the end of the century and into the next.

For the most part, till now spaceflights have been short ones, measured in days. What is needed next is something that will allow human beings to make a more-or-less permanent home in space.

The nearest thing for the purpose that the United States has had was

Skylab, which during 1973 and 1974 was occupied on three separate occasions by groups of three astronauts. On the third occasion,the astronauts remained in space for nearly three months.

The Soviets have had a number of vessels used for extended missions and one cosmonaut has remained in space for seven months. Another, on three separate flights, has logged a total of nearly a year in space.

These are only beginnings, however. What is needed is a true "space station," something more elaborate than anything either the United States or the Soviet Union has put up. It must be in high enough orbit to remain in space for an indefinite period, and it must be continually occupied.

Naturally, the occupants of a space station will not remain there indefinitely—at least not during the early years of such structures. Prolonged exposure to zero-gravity induces some distressing side-effects. Bones and muscles weaken, so that cosmonauts who have stayed in space for half a year or more have had some trouble reacclimating themselves to normal Earth conditions. A three-month stay, however, can be handled fairly well, particularly if the astronauts exercise regularly.

Plans are therefore underway now to have a space station in orbit about the Earth at a height of not less than three hundred miles above Earth's surface. The living and working quarters may consist of four or five large cylinders held together firmly. Such a space station should house eight astronauts under shirt-sleeve conditions, each of them serving a three-month shift and each being replaced when the shift is over so that the space station is continually occupied.

The value of such a space station is manifold. For one thing, long-range experiments can be carried out. Materials can be processed; welding and purification procedures can be tested out on a large scale and in detail.

Then, too, the Earth can be kept under continuous observation—and for the purpose the orbit of the space station would be tilted markedly to the equator (28.5 degrees is the degree of tilt often mentioned) so that the tropic zone and both temperate zones could be viewed with ease.

The space station will be large enough to store satellites for future launches and to have facilities for the servicing and maintenance of satellites already in orbit. The space requirements for all this will expand with time, of course, but the space station will be so designed that it can be expanded by hooking on additional units.

The space station can also serve as a base for the building of still larger structures that, when completed, would have been too large and massive to launch into orbit if they had been built on Earth's surface. Instead, the smaller components will be brought to some point in space which can be reached by astronauts from the space station, and there they will be put together.

Naturally, space stations will be expensive to maintain. Shuttles will frequently have to travel from Earth's surface to the space station and back in order to deliver those astronauts beginning their shift and take back others who have completed theirs. In addition, supplies will have to be delivered and wastes removed; disabled satellites will have to be brought in, repaired satellites taken out, and so on.

Any way in which service can be diminished, and in which the space station can make itself more self-supporting will, of course, reduce the expense. One obvious way of doing this is to have the space station collect the energy it needs from space.

This is not difficult, in principle. The station is bathed in sunshine during half its orbit about Earth; sunshine that is constant and is not absorbed, to a greater or lesser degree, by an atmosphere. The space station being planned now, therefore, will be equipped to exploit solar energy.

The space station will have an overall T-shape. A long girder will extend upward from the living and working quarters. Near the top of this girder, there will be a transverse structure forming the crossbar of the T. To each half of the bar there will be attached four solar arrays, two above and two below, making eight in all. These can be moved so as to face the Sun always.

Each solar array will contain banks of photoelectric cells that will convert sunlight into an electric current.

Such photoelectric cells are routinely used for some energy-gathering purposes on Earth right now, and even in space, though always in small-scale installations. The solar array used in connection with the space station would be a more ambitious and larger installation than anything we now have.

Even though the space station is expected to require ten times as much electricity as anything else we have put into space, the solar arrays that are being planned should be addequate. Not only would they meet the station's requirements, but they would produce an excess that could be used to charge storage batteries. During the portion of the orbit that passes through Earth's shadow, the stored electricity could be drawn on.

Another way of making a space station more nearly self-sufficient, and to cut down the transportation requirements from Earth's surface, involves the matter of food.

Until now, all human activity in space has involved spaceships that are in orbit about Earth or in transit to the Moon and back. When the duration of such flights is measured in weeks, it is possible for the ships to bring aboard, at the start, all the food that will be required for the flight and to find room in the craft to store any wastes that are produced. Those few spaceflights that last for months have invariably been in close orbit about the Earth so that, when necessary, new supplies could be brought up to them and stored wastes removed.

Eventually, though, when longer and more distant flights become necessary, such a procedure would be impractical. Ways of recycling must be evolved, and such matters are on the drawing board now.

When food is eaten, its components are combined with oxygen to form waste material, including carbon dioxide. In the process, energy is produced that serves to maintain the life of the eater. Plant life can be grown, making use of the carbon dioxide and the waste to form food and oxygen anew, provided sufficient energy, in the form of light, is supplied.

A greenhouse, then, will be built in conjunction with the space station. It will be lit by electricity drawn from the space station's solar array, and in it a variety of fruits and vegetables will be grown. Such a greenhouse might well supply fifty percent or more of the food eaten by the personnel on the station.

There is no reason (we can hope) why the facilities and capabilities of the space station cannot be open to all nations even if it is built (as would appear most likely, at the moment) by the United States alone. If the space station could be made into a global project, serving global needs, it could by its very existence help foster international cooperation and peace.

The space station will serve as a training ground for much more ambitious projects. For instance, the work it does in repairing, maintaining, and relaunching satellites may serve as a preparation for the construction and launching of new types of satellites designed to serve new purposes.

Thus, it is certain that scavenger satellites will have to be developed. These would be capable of removing the kind of space debris that has been steadily accumulating in orbit about Earth in the thirty years of the space age. Such debris naturally raises the spectre of possible collisions. Even small bits of matter, when moving at orbital speeds, can do measurable damage on collision. The debris, if not dealt with, may begin to limit space activity uncomfortably by the time the twenty-first century opens.

Then, too, all the work astronauts on the space station perform in carrying through industrial processes will teach humanity how best to make use of the special properties of space—zero-gravity, hard vacuum, energetic radiation from the Sun, and so on.

Special structures—factories, not to mince words—can be built in which these properties can be taken advantage of to produce, in quantity, such objects as electronic components and microchips. In such structures, purification and welding procedures can be carried through, too, as well as an unlimited quantity of other industrial steps limited only by human ingenuity on designing the necessary devices for the purpose.

What's more, the use of automated procedures and robots would make unnecessary the continued presence in these space stations of human beings, who might show up only occasionally for some necessary piece of repair or maintenance.

This might be the beginning of the transfer of much of Earth's industries into orbit. It would relieve Earth of some of industrialization's disadvantages, since some factories would be removed from Earth's surface, without depriving us of the advantages, since those same factories might not, after all, be more than a thousand miles away.

What's more, this would mark one of the points at which space might not only cease being a drain on Earth, might not only come to be merely self-supporting, but might actually produce an excess of products that can be shipped to Earth. Space will begin to support the Earth, rather than vice versa.

Naturally, people would want to be able to move to and from the various structures built in space. For the purpose, special vehicles are being planned, such as the "Orbital Transfer Vehicle," or "OTV."

The OTV will be reusable and space-based. Once built in space, it would be beyond the atmosphere and would already be moving in orbit with a speed approaching five miles per second. It would have to add only a small amount of additional speed to move into a higher orbit, even one considerably higher. For this reason it would require much less fuel than vessels that must start from Earth's surface, and it would be able to carry proportionately larger payloads.

The OTV, therefore, would greatly increase the ease and efficiency with which astronauts might reach the particular height of 22,300 miles above the surface of the Earth. At this height, an object would orbit the Earth in 24 hours, the same time it takes Earth to turn on its axis.

If a ship at this height were in a circular orbit and moved in the plane of Earth's equator, it would move about Earth's axis in lock step with Earth's surface, where someone looking upward would see the ship remaining directly overhead indefinitely. The ship would then be in "geostationary orbit," and it is *only* in that orbit that an object could be maintained over one spot on Earth's surface without the continual use of energy. (Such an orbit can also be referred to as a "Clarke orbit," after Arthur C. Clarke, the well-known science fiction writer who back in 1945 showed how useful such an orbit could be in connection with communications satellites.)

The combination of space stations and OTVs would make it possible to build structures in geostationary orbit. A series of communications satellites at different places in this orbit would mean that signals could be sent to the satellites, relayed, and sent back with less difficulty than if each satellite was moving relative to Earth's surface. Such stationary positions might also be useful for navigational satellites, weather satellites, Earth-resources satellites, and so on.

The most important function of this orbit, however, might be in connection with energy satellites. The solar arrays built for the space station would give engineers invaluable lessons in techniques and precautions that would make it easier to build still more magnificent structures.

It seems conceivable that solar arrays, miles across, might be built in geostationary orbit. Where such arrays on Earth's surface, or even on the space station, would on the average be in darkness for half of the time, an array in geostationary orbit would usually miss Earth's shadow as it turned (due to the tipping of Earth's axis). The arrays would enter Earth's shadow for only brief periods during the days in the neighborhood of each equinox. Such an array would be exposed to sunlight for ninety-eight percent of the time altogether.

Combined with this is the fact that solar arrays in space can receive sunlight across the entire spectrum of wavelengths, thanks to the absence of atmosphere. It is therefore estimated that an array in geostationary orbit will pick up sixty times as much energy as that same array on Earth's surface.

A series of such arrays spaced about the Earth could convert sunlight first to electricity then to microwaves. The microwaves can then be beamed to a receiving station on Earth (with added simplicity, since the array would be motionless with reference to the receiving station). At the receiving station, the microwaves could be reconverted to electricity and distributed over the world.

Again, we would find the current of economic value reversing, for in this respect, too, space would begin to support Earth, rather than vice versa. A major portion of Earth's energy needs might be met in this way, removing the need for our vanishing fossil fuels, and eliminating a major source of pollution.

Unfortunately, however, the geostationary orbit is a limited resource, for it is only 165,000 miles long, and only so much can be fitted into it. It will have to be exploited wisely and with forethought. Clearly, the nations of the world will have to collaborate on its efficient use, and all might expect to receive the benefit that comes from satellites located there. It would be another strong push in favor of international cooperation and peace.

Where does all the material come from to build the various structures in space? Earth's resources are being stretched thin for the needs of Earth's own population, and even if enough can be found for delivery to space, the act of delivery would be very expensive.

Fortunately, there is an enormous piece of real estate that we can reach, real estate that we already *have* reached with old-fashioned, non-reusable spaceships. It is the Moon.

Once we have space stations in orbit, it will be only a matter of time before we return to the Moon. This time it will not be for temporary visits, but to stay—at least in relays. Once we have a permanent lunar presence, we will be able to study the Moon in detail and use it as a stable, airless base on which to establish a huge astronomical observatory.

More than that, we can establish a mining station there. Detailed studies have been made concerning the gathering of ore from the lunar surface. It could be hurled into space electromagnetically (not difficult, since the Moon's escape velocity is not much more than one-fifth that of the Earth's), and there, in space, it could be smelted with the use of solar energy.

Lunar material can serve as a source of structural metals, cement, concrete, glass, and even oxygen. With the Sun supplying energy and the Moon supplying material, it will be possible for human beings to build a large number of space structures without calling upon Earth itself for excessive supplies of energy or matter. Three important elements the Moon cannot supply, however, are hydrogen, carbon, and nitrogen, but Earth has a large enough supply of these to fill all requirements for a long time.

One might argue that Earth would have to supply people, but even that is not necessarily so. Among the structures that will be built in space might be space stations so large that they could hold thousands of people, rather than a dozen; so large that rotation at a not-excessive speed would supply a centrifugal effect that would prove an adequate substitute for gravitation. So we could avoid the deleterious effects of zero-gravity.

In such stations, human beings might be able to live lifelong, generation after generation. We would then have new permanent habitats for human beings—not just on Earth's surface, or in airtight enclosures just below the lunar surface, but in numerous space settlements in the lunar orbit and even beyond.

The space settlements could control the quantity of sunshine they receive and would be free of bad weather. They would keep out deleterious life forms (at least to a greater extent than we can on Earth). Conditions on space settlements might thus prove ideal for farming. With the experience of the greenhouses attached to the original space stations behind them, the settlements could make use of adequate cycling procedures and minimal supplies of fertilizer from Earth to produce plants and even small animals in supplies far greater than they would themselves need. Thus, space would supply not only energy and manufactured devices to Earth, but food as well.

Furthermore, space settlements would offer an ideal inducement for space travel. At their distance from Earth the escape velocity would be very low. Between that and the omnipresent vacuum of space, fuel requirements would be moderate, and advanced methods of propulsion (ion-drive, solar wind sailing) might be made practical, saving additional quantities of fuel.

Furthermore, the space settlers would be far more psychologically suited than Earthmen to the undertaking of long flights. The space settlers would be more accustomed to space, and to living inside artificial structures on cycled food, water, and oxygen.

It is they who might make the routine flights to Mars and the asteroids by the mid-twenty-first century. The asteroids in particular would offer mining possibilities even beyond those of the Moon—and would supply the hydrogen, carbon, and nitrogen that the Moon does not.

By the end of the twenty-first century, humanity might in this way be ready to penetrate the vast spaces of the outer Solar System, and, eventually, to move even beyond—to the nearer stars.

── 57 ──

LIVING ON THE MOON

The next two essays were written in 1969 before the first landing on the Moon. I include them as examples of predictions that have been overtaken and, to some extent, falsified by events. (I can easily make myself seem more prescient than I am by selecting predictions that seem to be coming true, while suppressing those that have been falsified—but that would scarcely be honest.)

In these two essays, I assume that there is water to be obtained from lunar rocks, and that there would even be organic materials in the lunar soil. Sorry— while the suppositions were fair ones at the time, they happened to be wrong.

A study of the Moon rocks brought back by the astronauts shows no signs of water. The Moon not only seems to be bone-dry, but seems to have been bone-dry all along. What's more, there are no organic compounds we can locate.

This doesn't make total nonsense out of the essays, however. It merely means that lunar colonies will have to import all water and organic materials from Earth to begin with—which I actually say in the essays anyway. It's just that later on they'll have to develop sources from places other than the Moon itself—from Mars, from the asteroids, from the occasional comet that may pass nearby.

PART I: WATER AND AIR

As they circled the Moon, the Apollo 8 astronauts reported it to be a vast deserted wasteland, and returned happily to the beautiful oasis of Earth.

Yet they told scientists nothing the scientists didn't already know. For three hundred years, astronomers have known that the Moon was a dead world, without air, without water. It was a changeless desert, infinitely worse than the Sahara, baked for two weeks under a harsh Sun, frozen for two weeks in a sub-Arctic night.

Yet there are people who talk of establishing a colony on the Moon. Is there any sense to that?

Of course there is. A permanent colony on the Moon would carry through numerous lines of research that could prove of the greatest value.

By exploring the Moon's crust, the colonists could learn more about the early history of the Solar System and the early history of Earth itself. (Our own crust has been constantly altered by the action of wind, water, and living organisms; the Moon's crust is much more primitive and untouched.)

Scientists on the Moon could study the organic compounds in the lunar crust and determine how these have been formed by the action of the Sun. This could teach them a good deal about the development of life on the Earth and even clarify some of the fundamentals of life-chemistry. (It is conceivable that we may be put considerably further on the road to the cure to cancer by a study of the Moon's crust—which might surprise people who are always asking, "Why don't we spend the money on cancer research instead of on rockets to the Moon?")

Factories on the Moon could use the huge stretches of vacuum for specialized technology. They could use solar radiation, too, and the frigid temperatures available in the lunar night. Small electronic devices and complex chemicals may be easily prepared on the Moon, which could be prepared only with difficulty, or not at all, on the Earth.

An observatory on the Moon could explore the Universe through telescopes that are not affected by the dust, clouds, and glare of the Earth's atmosphere.

A colony on the Moon could . . .

But wait! No matter how many wonderful things a group of Moon explorers might do if there were a colony on the Moon, what good is there in even thinking about it? How can a colony on the Moon be supported? How can people live on that forbidding world? Why dream an impossible dream?

Because it isn't impossible. There is every reason to think that men can live very comfortably indeed on the Moon, once the initial investment is made.

Life on the *surface* of the Moon would be extremely difficult, to be sure. But then that is not what we need visualize; the life would be underground. Trips to the surface would be possible; exploratory trips across the surface, with proper protection against radiation and against temperature change, would also be possible. But for day-to-day living, the colonists would have to remain underground.

All the obvious difficulties of the Moon would vanish if there were a few yards of lunar crust between the colony and the surface. During the Moon's long two-week day, the surface temperature reaches the boiling point of water in certain spots. During the Moon's long two-week night, the surface temperature drops to something like –200° F. No one would want to face either situation, and the colonists in the underground cavern would not have to.

The lunar crust is a poor conductor of heat. Very little heat would leak into the underground cavern during the day; very little would leak out at night. The temperature in the cavern would be even, day and night, and might be quite close to what we are accustomed to on the Earth. If it should be rather on the cool side, a little heating would take care of that.

On the surface of the Moon, there is deadly danger from the radiation of the Sun and from a continuing rain of tiny meteors. On Earth, a thick layer of air absorbs the hard ultraviolet and x-ray radiation of the Sun and burns up the meteors as they fall. On the Moon, there is no blanket of air to save the colonists—but a few yards of lunar crust would do the job perfectly well.

In an underground cavern there would be no dangerous radiation from the Sun, no meteors.

Cosmic rays will reach the cavern, to be sure, but they reach us on Earth, too. A giant meteor might smash its way into the cavern, but a giant meteor might destroy a city on Earth, too.

Then, too, the underground Moon colony (let's call it Luna City) would avoid some of our troubles. It would not be subjected to any of the vagaries of our weather. It would not be subjected to any of our animal pests (unless colonists carelessly brought them along). With some care to begin with, Luna City might even be free of the more vicious kinds of diseases and germs.

Perhaps you wonder about the psychological effects of living in a cave. People will manage.

Think how far removed huge cities like New York or Tokyo are from the kind of life in which mankind evolved—yet men manage. In fact, there will be little difference between the environment of an underground Moon colony and that of the huge office/apartment building caverns in which millions of New Yorkers spend ninety percent of their time.

Naturally, Luna City won't be established at once. The first explorers will touch down on the Moon, then leave after making some tests and collecting some samples. On a later trip, a ship may be partly submerged in the lunar crust and will serve as home-base for a couple of Moon explorers who may then stay for several days before being lifted off.

Such ship-based explorers can make a beginning in the digging out of a cavern. Little by little, the cavern will be made larger and more elaborate—until Luna City becomes a tiny metropolis.

But still, how will the men of Luna City be supported? Where will they get their water, their air, their food?

Originally, of course, they will get it from Earth. *Everything* will be transported from Earth. Each ship that brings men to the Moon will also bring what supplies it can of food, water, and air.

This, however, can and must be only a temporary expedient. The long, fragile pipeline from Earth to the Moon could too easily be broken. There would be too many reasons for the kind of delay that might place the Moon colony on the edge of extinction—or over that edge.

Besides, it would be too expensive. As long as everything was being brought from Earth, Luna City would continue to be a severe drain on Earth's resources, the kind of drain which humanity might not be willing to tolerate for long.

The aim, then, would be to make Luna City "ecologically independent." That is, it must be able to get all its essentials from the Moon itself.

But how could this be done? Where on Earth—or rather, where on the Moon—could the men of Luna City obtain their water, for instance?

When astronomers agree that there is "no water" on the Moon, what they really mean is that there are no open stretches of liquid surface water there. There are no lunar oceans, lakes, or rivers.

It is almost certain, though, that there is water on the Moon in other forms.

Water is a compound. That is, its molecules are built up of different kinds of atoms. In the case of water, the atoms involved are those of hydrogen and oxygen. Hydrogen atoms are the most common in the Universe, helium atoms are second, and oxygen atoms are third. Helium atoms don't form compounds, however.

This means that of all the compounds in the Universe, water must be most common. Fragments of water molecules have even been located in outer space.

Any astronomical body that is reasonably large and reasonably cool must at some time have included large quantities of water. The Moon is a little too small, however, and its gravity is a little too weak, to have kept water on its surface. All that surface water has disappeared in the billions of years of the Moon's existence.

In its youth, though, before the surface water had time to go, the Moon must have had a thin atmosphere and may have been a sparkling world of lakes and rivers. In fact, satellites which have approached and circled the Moon have sent back photographs which show twisting channels gouged into the Moon's surface that look *exactly* like the dry beds of ancient rivers.

It is almost certain, then, that though the surface water is gone, there must remain quantities of water underground. It may be that there are underground veins of ice that might be mined as coal is on Earth. Failing that, it is sure to be found that water is loosely combined with the lunar rocks themselves.

If such rocks are baked, water vapor will be released, and this can be condensed to liquid water. Whether from ice or from the rocks, there seems every reason to suppose that enough water can easily be obtained to support Luna City.

Mind you, you mustn't think of water in Earthly terms. Mankind has spent so many thousands of years on a world with vast oceans of water that he has never learned to use water properly. He has always wasted it carelessly. Lakes, rivers, and the ocean itself are now used as sinks into which to throw deadly poisons. It is only in recent years that we are beginning to realize the danger to which all life is being subjected by this criminal carelessness.

In Luna City, of course, water would be used with extreme care. What's more, it would be constantly recycled. All waste water, including urine and sewage, would be distilled, and the fresh water so obtained would be reused. (The Soviet Union has recently reported that three men were kept in a simulated space cabin for twelve full months and made use of such recycled water.) It may seem horrible to drink water distilled out of wastes, but this happens constantly on Earth. All the fresh rain, all the fresh mountain streams, consist of water distilled out of the ocean, out of wastes, out of all sorts of undrinkable material, by the action of the Sun.

Naturally, recycling cannot be done with perfect efficiency. There is always some wastage, and it is only this wastage (a small fraction of the total water used) that would have to be replaced by mining ice-veins or baking rock.

But blasting through the ice-veins, or baking rock, takes energy. Where does the energy come from for that—and for lighting and heating Luna City, and for keeping all its machinery running?

We might suppose that electricity will be supplied in ample amounts by a uranium fission reactor that will be brought from Earth piece by piece and put together on the Moon.

Once that initial investment is made, Luna City will turn to the Moon's crust for uranium supplies to support the reactor. Or perhaps by the time Luna City is established, practical hydrogen fusion reactors may be devised.

Fusion reactors can be expected to deliver many times as much energy per weight of fuel as ordinary uranium fission reactors would. What's more, fusion reactors use heavy hydrogen as their fuel, and this can be obtained from water.

But we don't have to count on the uncertain chance that there will be enough uranium in the crust around Luna City to support a fission reactor. Nor do we have to count on the uncertain chance that a practical fusion reactor will be devised in time for Luna City. There is an energy supply on the Moon right now, in unlimited quantities and free for the taking.

The energy is in the Sun's radiation. We can develop solar energy right here on Earth, for instance, by spreading banks and banks of appropriate devices where sunlight can strike them. These devices, called solar batteries, can convert the Sun's radiation into electricity directly.

On Earth, the Sun is frequently obscured, particularly in the dust and smog of a city where the energy is needed most, and most particularly in the winter when it is needed most.

On the Moon, though, solar batteries can be spread out over many square miles of the Moon's empty surface. The Sun, which shines for two weeks at a time over any particular spot on the Moon, without ever being obscured by clouds, will pour copious quantities of energy over the batteries. It would be easy to arrange to have enough energy absorbed and stored to support Luna City—not only during the day, but also during the two-week night when solar energy is absent.

Once Luna City has water and energy, it doesn't have to worry about air. Water, properly treated with an electric current (the process is called "electrolysis"), can be broken up into oxygen and hydrogen. The oxygen can be used to keep the air breathable, the hydrogen for other purposes.

Nor need we use oxygen as wastefully as we do on Earth. It, too, can be carefully recycled. Additional supplies of water will have to be electrolyzed only to make up for the imperfection of recycling.

But then what about food? Our Moon men can drink and breathe, but what can they eat? We'll consider that in the second half of this article.

58

LIVING ON THE MOON

PART II: FOOD

In my last essay I pictured a permanent colony on the Moon—Luna City—which derived its energy, water, and air from the Moon itself. The question now is: How will it manage where food is concerned?

To begin with, of course, our Moon explorers will bring their food from Earth. Even after they have established themselves in an underground colony and have developed their own lunar supply of energy, water, and air, there will be a tendency to continue to depend on Earth for food.

For one thing, energy is energy, water is water, and air is air. It doesn't matter whether they are obtained from the Moon or the Earth.

Food, however, comes in a thousand varieties of flavors, colors, textures, and odors. Some foods are hard, some soft, some hot, some cold, some crunchy, some smooth, some spicy, some bland. The Moon could not conceivably supply food in such variety for a long time, and the tendency would be to depend on Earth.

Luna City won't be driven as hard to produce its own food as it will its own water, air, and energy. After all, one can skimp on food at need (though this is never pleasant) longer than one can skimp on water and air, and the terror of a delayed arrival of a supply ship would not be so intense as long as water, air, and energy were in ample supply.

Furthermore, if the incoming spaceships are not required to carry water or air for the colony, they can be more liberal about bringing food.

Yet the food a spaceship can bring must be carefully selected. It would be the height of folly, for example, to bring over whole turkeys, or legs of lamb with the bone left in. The weight of inedible portions would be a dead loss, and if one figures out how many pounds of fuel must be burned in order to transfer one pound of mass from Earth to Moon, a pound of uneatable lamb-bone becomes an expensive proposition indeed.

The tendency would be to bring food prepared in such a way that it is completely edible. Turkey white meat rather than whole turkey; chopped meat rather than porterhouse steak. As the population of Luna City becomes larger, the pressure for concentration of food will increase.

One can easily imagine the extreme. One can prepare semisolid mixtures of fats, carbohydrates, and proteins, packaged in a crisp, edible crackerlike container, with all necessary vitamins and minerals added, and a variety of flavors, too. Perhaps twenty ounces per day per man would supply the food requirements of Luna City, providing always that Luna City could itself supply all the necessary water.

But even this builds up. Suppose Luna City has grown to have a population of one hundred (surely a modest number) and that a supply ship comes from Earth every month. To bring ample supplies of this economical food concentrate to everyone in the city, it will be necessary to carry three tons of it. You can see, then, that it would be a great relief to the men who must ferry these ships to the Moon if Luna City were to take over some of the responsibility for its own food supply.

Here are some of the factors that would cause Luna City to strive to do so. Suppose it does somehow grow its own food, enough to supply ninety percent of its needs. That ninety percent might be rather poor, but if the ship need only bring six hundred pounds rather than three tons, it might be willing to bring along such foodstuffs as lamb chops and apple pie. Even a few of such things would be better than none at all.

Were Luna City to produce its own food, the colony would become far more ecologically independent. Water and air must be recycled if the colony is to survive; that is, wastes must be turned back into drinkable water and carbon dioxide must be changed back to oxygen.

This can be done by purely technological methods—distillation, chemical treatment, and so on—which will require large expenditures of energy. It might, on the other hand, be done automatically by the very procedures that produce the food. With automatic recycling, Luna City will become a miniature Earth. Of course, it will have a far smaller margin for human or technological error—and such transport and recycling operations will have to be tightly supervised.

But what kind of food ought Luna City produce? Naturally, large animals like cattle are out of the question. Even a supply of newborn calves would be difficult to ship out to the Moon. It would be much easier to send a crate of fertilized eggs to the Moon, even hatching them out in transit. But then, what do the chicks do on the Moon. What do *they* eat?

It will be necessary, at the start, to cultivate food that does not, in its turn, require food. We need an edible organism that grows on inorganic material, and that means plants. As far as the Moon-grown food is concerned, the early colonists of Luna City will simply have to be vegetarians and depend on supply ships from Earth for any meaty snacks.

It is plant life, furthermore, that will make it possible to recycle carbon dioxide and wastes. Ordinary green plants absorb carbon dioxide from the air and combine it with water to make the carbohydrates, fats, and proteins of their tissues. In the process, oxygen is left over and is discharged into the air.

With plants growing in Luna City, men can breathe in oxygen and breathe out carbon dioxide freely, for the plants will use the carbon dioxide to produce food and will restore oxygen to the atmosphere. Then, too, properly sterilized and treated human wastes will serve as excellent fertilizer for the plants (and make it less necessary to process Luna rock for chemical fertilizers). In this way, the wastes will be converted back into food.

Men release energy by turning oxygen and food into carbon dioxide and wastes. If plants are to turn carbon dioxide and wastes back into oxygen and food, that released energy must be restored. On Earth the source of the necessary energy is sunlight. Plants use the energy of sunlight to manufacture food and oxygen out of carbon dioxide, water, and minerals.

On the Moon, of course, sunlight is present in plenty. The solar batteries spread out over square miles of the lunar surface would produce all the needed electricity, and some of this can be used to light fluorescent bulbs which would supply the plants with the visible radiation they need.

With careful tending, men, plants, and energy could be maintained in good balance as the colony grows.

What kind of plants would Luna City grow? Clearly, the chief requirements would be proteins and calories. It would be foolish to grow plants with an eye to vitamins and minerals, for these are only needed in small quantities and can be added to the basic diet, if necessary. The Moon's crust has an almost infinite store of minerals, and any needed supplies of vitamins could be brought from Earth by rocket without undue expense.

It would also scarcely pay the colony to grow those plants which would include much in the way of inedible portions. Inedible cobs, stalks, roots, or leaves would be an unbearable waste.

If Luna City were interested in a maximum of edible fat, protein, and carbohydrate, and a minimum of waste, it would have to turn to the world of microorganisms and grow algae, yeast, and bacteria. The algae would use light energy, and the yeast and bacteria would live on simple energy-yielding chemicals obtained from the algae. (Some strains of bacteria might live, in part, on hydrogen gas that could be obtained from the electrolysis of water.)

Microorganisms grow exceedingly fast. Nothing could better them in pounds of edible material turned out in unit time. There would be no roots, stems, or other inedible portions to speak of. They would be nothing but food.

The psychological difficulties involved in eating such microorganisms would be easily overcome. When one thinks of what man has learned to eat, or even considers delicacies (raw oysters, caviar, and squid, to name just a few, and say nothing of rattlesnake and chocolate-covered grasshoppers), an algae-cake would be easy.

There would, however, be a question of flavor and texture, or the lack of it. But then, when necessity drives. . . .

It is quite likely that one of the most ardent pushes in youthful Luna City would be that of attempting to adjust the taste and quality of the native diet. A great deal of effort and imagination will be put into preparing the microorganismic base in such a way that, in addition to being nourishing, it will also be tasty.

The use of synthetic flavors and physical treatment will produce a host of imitations: from mock-soup to mock-nuts, with mock-steak, mock-potatoes, and mock-salad in between.

I imagine also that the farmers of Luna City (who will have to be competent biochemists, geneticists, and microbiologists in order to be farmers) will be working constantly with new strains. By exposing ordinary strains briefly to direct sunlight on the Moon's surface, radiation-induced mutations will be produced. Most will be worthless, but occasionally a new strain will bring with it a particularly interesting taste.

It is not beyond the realm of possibility, in fact, that in the end, a lunar cuisine would be developed which would be quite different from any of the Earthly types, and would develop properties all its own.

For instance, we are dealing with tiny cells that don't possess the kind of tissues needed to bind together the parts of large plants or animals, so there would be no need for prolonged cooking. Microwave radiation could probably do what is needed in a matter of a minute.

Then, too, in suspension, the food could be piped to individual homes. We can imagine (eventually) a small home in Luna City in which a kitchen consists of a unit through which rationed quantities of various strains of microorganisms can be delivered. The push of a button will select one strain or another, or a combination. Another button will control the microwave unit and the compression unit, so that the food will be delivered in free suspension as a soup, thickened into paste, compressed into cake form, heated into a wafer, or even cooled into a kind of dessert.

Every once in a while a new strain might be announced, and this could be a time of considerable excitement, as families get together to sample the new item and then compare notes.

The changes and switches that can be produced will be limited only by imagination, and in the end the lunar cuisine may come to be highly valued for its novelty by those who don't have to live on it every day. Gourmets from Earth might visit Luna City to sample its food as well as to see the sights.

This is not to say that Luna City will be condemned to live on microorganisms forever. As the city grows and its economy expands and diversifies, it will have greater margin for frivolity.

There will come a time when, instead of growing food only in illuminated or chemically-fed tanks, the soil of the Moon itself can be used for growing limited quantities of ordinary crops.

Naturally, such crops could not be exposed to the surface conditions as they are in their unmodified state. The heat, cold, vacuum, and solar radiation would be too much for any of Earth's higher plants.

The crops would be grown under glass in a thin atmosphere rich in carbon dioxide. The glass would be so designed as to block hard radiation, while letting visible light through. (The glass would probably be gradually scoured by tiny, dust-sized meteors, and would require bit-by-bit replacement.)

To take care of the accumulating heat of the two-week daylight period, there could be a double-glass dome enclosing circulating water. The heated water, stored below, would serve as a heat-reservoir that would circulate again to help keep the dome warm during the increasing frigidity of the long two-week visit. (The dome would be intermittently lighted, artificially, to keep the plants growing through that long night.)

With time, acres and acres of the Moon's surface could be turned over to this domed farming. Because of the absolute predictability of the weather on the Moon, and because of the careful control (chemical and otherwise) exercised by the farmers, the plants would grow with much greater efficiency than on Earth.

Not only could such normal crops supply the Luna City colonists with a perhaps welcome change from the native Moon-food, but the relatively inedible portions of those plants could be used to feed a limited supply of small animals—chickens, rabbits, and so on—which could, in turn, supply the colonists with occasional meals of real meat.

Indeed, as Luna City developed, grew, and strengthened, it would become more and more like an Earth city—always excepting the one factor of the environment which the colonists could neither adjust or modify: the gravity. (The surface gravity on the Moon is just one-sixth what it is on Earth, but that is another story.)

By 2100, I supsect, there will be people on the Moon who will be quite disgusted with the "decadence" of the world and who will long for the sturdy virtues of their pioneer grandfathers.

This disgust, mixed with some adventuresomeness, may well lead a migration to the newly established pioneer colony called Mars City, off on the planet Mars.

59

THE SKIES OF LUNA

1969: Man reaches the Moon.

1989: Permanent colony established on the Moon. [*This essay was written in 1971, and turned out to be a bit optimistic on this point.*]

2009: Commercial company opens shuttle craft to Luna City. Prices reasonable. Tourists invited.

So who wants to go? What's there to see on the Moon? Miles and miles of dirty beach? The fenced-off site where Neil Armstrong's footprints will be visible in perpetuity? The original American flag? Two weeks inside a dome made up to resemble an American apartment house?

Is that all?

No, it isn't all! Far from all! There's a show put on by the skies of Luna that's nothing like anything ever seen here on Earth. A planetarium might put on an imitation of it, but it would be like a wax orange compared to the real thing.

Just to be on the Moon and take one good look at its night sky—even with a space helmet on, or from behind the protective glass of a lunar observatory—will reveal at once that what we call a night sky on Earth is nothing but a pale, washed-out substitute.

There are no clouds on the Moon, no mists, no fog, no city lights; nothing to hide the stars, nothing to dim them, nothing to drown them out.

Better than that, there's no atmosphere on the Moon. On Earth, even during the clearest, coldest night, deep in the desert and far from any city smog and light, the air still absorbs thirty percent of the starlight that impinges upon it. If you look toward the horizon, you look through the air on a slant and therefore through a lot of extra thickness. Stars that manage to shine through the air from zenith, and survive just enough to be dimly seen, blank out as they approach the horizon.

On the Moon there is no atmosphere to absorb starlight, so that each star shines nearly half a magnitude brighter than it does on Earth, and keeps its brightness from zenith all the way to horizon. This means that many stars just under the threshold of visibility as seen from Earth grow bright enough to be clearly visible in the skies of the Moon.

The unaided 20/20 eye on Earth, under the very best conditions, can make out perhaps 2,500 stars in the sky at any one time. On the Moon, the same eye would make out nearly 6,000 stars.

The bright stars would all be there in the Moon's sky, brighter than ever, and making up the familiar constellations. They would be set off, however, by hordes of dim stars not seen on Earth by the naked eye, and would lend those constellations a richness undreamed of here.

What's more, the stars of the lunar sky wouldn't twinkle. It is the temperature differences in air that bend the tiny starlight-beams this way and that, making the little sparks constantly shift position. This gives the stars a soft beauty but it wearies the eye. On the airless Moon, each star, however faint it might be, is fixed. The eye can follow the patterns of the constellations; the mind can create associations and pictures in those patterns in endless variations.

And there will be ample time to study those patterns, too, for the lunar night lasts 348 hours—or just over two weeks.

Then the Sun rises in the east. It's the same Sun we see from Earth and appears to be the same size—but it's a lot more dangerous. There's no air to soften the radiation or absorb the x-rays and far ultraviolet rays before they reach us. There is no ocean to absorb the heat and keep the temperature rise moderate. There are no winds and currents to spread the heat. At the lunar equator, the temperature at noon reaches that of the boiling point of water.

But that's just on the surface, of course. In Luna City, underground, the temperatures will always be moderate. And through television cameras, one can still watch the sky.

On Earth, the daytime sky is a featureless blue—except when it is obscured by clouds—because the air scatters light. The short waves (green, blue, violet) are scattered to a greater extent than the long ones (yellow, orange, red) so the sky is blue and the Sun seems faintly golden. The stars blank out against that scattered light; they are still there but invisible. Even the Moon can scarcely be seen when it is shining in daylight.

On the Moon, where there is no air to scatter the light, the daytime sky remains black, utterly black. The Sun's light, reflected brightly from the Moon's crunchy surface, would dazzle the eye and limit the clarity with which one could see the stars. But suppose you watch from inside the dome, with the television cameras turned to a section of the sky. The brilliant ground would be invisible, and if the Sun itself were not in the field of vision, the stars would be as visible by day as by night.

The Sun itself would be a rare sight to behold—though not directly of course. It is easy to imagine an opaque region on the television screen just large enough to cover the shining disc of the Sun. Suppose this opaque region were an exact fit and were shifted by a timing mechanism so that it continued to stay in front of the Sun as the body moved slowly across the heavens.

On Earth, this would make no difference. The atmosphere would remain full of light; the sky would remain blue; the stars would remain invisible.

On the Moon, however, with no air to scatter light, hiding the disc of the Sun would give the effect of having no Sun at all in the sky. Except that if the opaque region were an *exact* fit, the bright red rim of its atmosphere would be visible. The corona would be seen. The Sun would be in total eclipse for as long as the opaque cover is maintained, with all the beauty that can be seen only so rarely on Earth.

So far, though, it might seem that the Moon's sky offers only what Earth's does, only more of it. The Moon's sky is blacker for longer, the Moon's stars are brighter and more numerous, the Moon's Sun can be put into perpetual eclipse.

If we were to go up in a stratospheric balloon to a height of twenty miles above the Earth's surface, some ninety-eight percent of the Earth's atmosphere would be below us, and what we would see in the sky would then be almost what we could see from the Moon. Is the quarter-million-mile trip to our satellite necessary, then?

Why not, if the Moon's sky offers us something we couldn't see at all from the Earth under any conditions—and something so remarkable it is worth travelling the distance for?

In the Moon's sky there is the Earth! The Earth is, so to speak, the Moon's "Moon."

But what a difference! The Earth is nearly four times as wide as the Moon, and so it appears in the Moon's sky, four times as wide as the Moon does in ours.

The brightness of the object in the sky depends not on its width, but, all things being equal, its area, which is the square of the width. The Earth is 3¾ times as wide as the Moon, so the Earth in the Moon's sky has 3¾ x 3¾ or 14 times the area of the Moon in our sky. All things being equal, the Moon's Earth ought to be 14 times as bright as the Earth's Moon.

But all things are not equal. The Moon's visible surface is bare rock that absorbs most of the sunlight that falls upon it. Only about seven percent of the sunlight falling on the lunar surface is reflected back into space. The Earth, on the other hand, has an atmosphere which is more or less filled with clouds, and these are much better mirrors than bare rock is. Some twenty-eight percent of the sunlight that falls on the Earth's atmosphere is reflected back into space.

The Earth, square mile for square mile, reflects four times as much sunlight as the Moon does. Combine this with the Earth's greater visible area and it turns out that the Earth would be fifty-six times as bright as the Moon. Remember, though, that there is no atmosphere on the Moon to absorb the Earthlight. Adding thirty percent for that reason allows us to end with the fact that the Earth as seen from the Moon is just about eighty times as bright as the Moon is seen from Earth.

If the brightness of the stars, as seen from the Moon, offers a romantic and beautiful sight, what are we to say to the large "Moon" presented by Earth, and to the brilliance of Earthlight?

The Earth, as seen from the Moon, passes through phases, just as the Moon does, as seen from the Earth—and in the same period of time. It takes just twenty-nine and a half days to go from "new Earth" through "half Earth" to "full Earth," back to "half Earth," and finally to another "new Earth" at last.

The phases of the Earth are exactly opposed to those of the Moon, however. When it is the time of the new Moon on Earth, it is the time of the full Earth on the Moon, and vice versa.

When it is full Earth on the Moon, the Earthlight is at maximum—eighty times as bright as the brightest full Moon any Earthman has ever seen from the surface of our planet. Under the cool brightness of the full circle of Earth in the Moon's sky, the lunar landscape is lit by light without heat, throwing the surface into soft highlights surrounded by black shadows—in imitation of sunlight but without any of the latter's harsh and dangerous effects and with a glow nothing on Earth can duplicate.

Nor is this just something we need imagine; we can *see* the Earthlight on the Moon. Earth is full (as seen from the Moon), and its light is at the brightest, when the Moon is new and visible from the Earth only as a thin crescent, if at all. When the crescent is thick enough to linger in the sky for an hour or so after sunset, so that the sky is dark, we can see beyond the pale crescent the faint outlines of the rest of the Moon, lit by Earthlight.

Because a brightly lit object tends to look larger than it really is (a well-known optical illusion called "irradiation") the Sunlit crescent of the Moon looks a little wider than the Earthlit main body. The crescent seems to enclose the rest, and the effect is known as "the old Moon in the new Moon's arms." It was considered an ill omen in past time.*

The "ill omen" thought is merely human suspicion. We should rather take the sight as visible evidence of the glory of the Earth as seen from the Moon.

Since the Moon always presents nearly the same face to the Earth as it circles us, the Earth seems nearly motionless in the Moon's sky. If we were standing on some point on the Moon near the center of its face (as seen from the Earth), the Earth would appear directly overhead and would more or less stay there. If we were to stand north of the central point on the Moon's face, the Earth would appear south of the zenith. The farther north we were standing, the farther south the Earth would appear. If we were standing east of the central point, the Earth would appear west, and so on.

But wherever the Earth appeared, there it would stay, and through the month we could watch its phases change from new to full and back again.

Nor is it only the slow phase-change we could watch. The Earth's face is far more variegated than the Moon's is. The Moon presents us only one face forever, and that face is unbroken by water, untroubled by air. There is nothing to disrupt the smooth expanse of light, except for the darker "seas" that form

---

* In "The Ballad of Sir Patrick Spens," there is the following quatrain:

> Last night I saw the new Moon,
> With the old Moon in her arm,
> And I do fear Sir Patrick Spens,
> Will surely come to harm.

the faint blotches on the Moon's white surface.

Not so the Earth. On Earth's face there are the ever-shifting clouds, forming their curling faint-blue patterns, patterns that are never quite the same. And through the clouds one can occasionally catch glimpses of the deeper blue of ocean, the faint tawniness of desert, the touch of mild green that is the evidence of life. Occasionally, the outline of a continent might be made out. Those parts of the outline most often seen would be the desert areas where clouds are few—the bulge of African Sahara, the polygon of Arabia, the curve of Autralia or the Chilean coastline, the main extent of Baja California.

The Earth rotates, too, once in twenty-four hours, so that each part is presented to the eyes of the Moon tourist.

Because the variations are endless, the interest can never fail.

In fact, the Earth's phases, its rotation, its cloud cover, its color, and its continental outlines don't exhaust all there is to watch. . . .

Additional interest arises out of the fact that the Earth does not hang *quite* motionlessly in the sky. It would, if the Moon's orbit about the Earth were an exact circle and if it moved in that orbit at a constant speed.

This, however, is not so. The Moon moves about the Earth in an ellipse. It is a little farther from the Earth (and moves more slowly) at some parts of its orbit than others. Without going into detail to explain why it should be so, this uneven speed results in the Moon not presenting quite the same face to us at all times. During part of its orbit, it turns a little so we can see just a small way beyond its eastern edge, and during the rest of its orbit it slowly swings back so that we can see just a small way beyond its western edge. This is called the Moon's "libration."

The effect of libration to someone standing on the surface of the Moon is to make the Earth swing back and forth about its average position in the sky over a one-month period. Under some conditions it would shift as much as sixteen degrees this way or that. This means that if its average position were at the zenith, it could shift one sixth of the way toward the horizon before swinging back.

This shift in position would not be very spectacular to the casual observer if the Moon were high in the sky, but suppose it were low in the sky.

Suppose a tourist on the Moon were standing near the eastern (or western) edge of the face of the Moon turned toward us. If we imagine ourselves watching this tourist through a telescope, we would see the Moon's libration carry him beyond the visible edge and then back again, over and over.

What the tourist on the Moon would see would be the huge globe of the Earth sinking toward the horizon, then vanishing below it, then rising above it eventually, only to begin sinking again—over and over.

The exact detail of the effect would depend on the exact position of the tourist. If he were in such a place that the average position of the Earth were somewhat above the horizon, it would sink in such a way as to skim the horizon and rise again, doing that horizontal skim once a month.

If he were farther east or west, so that the Earth was still lower in the sky, part of its globe would sink below the horizon once a month and then emerge—farther still, the Earth would disappear completely before rising again.

If the Earth's average position were exactly at the horizon, then for days on end it would be invisible. Once a month there would be a huge, slow "Earthrise," in which it would take two full Earth-days for the globe to rise fully above the horizon after its first appearance; and once a month there would be an equally huge, slow "Earthset."

Suppose the tourist were still further east or west so that only the extreme swing of the Moon's libration would bring him forward far enough to be just visible at the edge to a man on Earth. The effect on the Moon would be this . . .

Once every month, at some point on the horizon, there would be the slow appearance of a patch of light just edging its way above the broken skyline. Mountain tops at the opposite point of the horizon would shine dimly in that light. And then as slowly as it appeared, it would disappear, the entire effect lasting perhaps an hour or so. It would be a corner of the Earth and its light just barely being brought into view by libration.

Depending on the position of the Sun, the temporary appearance of the edge of the Earth would come at a particular phase and there might be the solid edge of full Earth or the narrow crescent-tip of new Earth.

Across about forty-one percent of the total surface of the Moon, the Earth is totally invisible at all times. (This is the "hidden side" of the Moon; the side we never see even at extreme libration, and which we knew nothing about until rockets carried cameras all around the Moon in the early 1960s.)

It might seem that tourist hotels are not likely to be built there, for the Earth is by all odds the most fascinating object in the Lunar skies and yet, who knows . . . ?

The Earth's giant globe distracts attention from the stars, and its light, reflected from the Moon's surface, would make the dim star-points less impressive. There may be travellers who would be content to be away from the garish globe of Earth, and from the crowds of tourists watching it, in order that they may be in relative isolation with the stars. Tastes differ!

But there remains one spectacle involving the Earth that even the most isolationist of tourists would surely be grieved to miss. That involves the combination of Earth and Sun.

The Sun, as seen from the Moon, moves across the sky more slowly than Earth's Sun does, for the Moon rotates about its axis only once in twenty-nine Earth days. The Sun rises, spends fourteen days crossing the sky, then sets, and spends fourteen more days making its way back to the point of sunrise again.

As the Sun crosses the sky, its distance from the Earth changes (assuming we are on the side of the Moon which has the Earth in its sky). The exact phase of the Earth depends on its apparent distance from the Sun in the sky. When the Sun is at the opposite point in the sky from the Earth, the result is full Earth.

Usually, when the Earth is in the sky, this means that the Sun is below the horizon, and full Earth takes place in a sunless sky. (The situation is precisely the same with full Moon on Earth.)

Suppose, though, the Earth is at the western horizon. Then, when the Sun rises in the east, there is full Earth in the west. As the Sun mounts higher, the Earth-phase narrows and the shadow creeps across its sphere. By the time the Sun is at zenith (a week after it rises), the Earth is half Earth, the lit half being toward the Sun. As the Sun descends toward the west, the Earth becomes a narrower and narrower crescent, the horns of the crescent always facing away from the Sun.

If the Earth were at zenith, there would be a half Earth, with the lighted half at the east at sunrise. This would narrow to a crescent which would thin almost to disappearance when the sun was near zenith and would start to thicken again as the Sun moved westward till it was half Earth again, the lighted half on the west, at sunset.

But what would happen when Sun and Earth were in the same part of the sky? Would the apparent distance between them decrease to zero?

Not necessarily.

The path followed by the Sun in the Moon's sky is such that ordinarily it passes either above or below the Earth. The Earth's narrow crescent shifts position from east to west, around the northern or southern edge of the Earth.

The amount by which the Sun misses the Earth's disc as it crosses from east to west varies. If it begins by missing it quite a bit on one side, it will move closer at the next pass and finally start missing it on the other side, farther each pass, till it starts moving back again.

Every once in a while, in the process of passing first on this side, then that, the Sun manages to make a direct hit, so to speak, and passes behind the Earth's disc. When that happens, sunlight cannot fall on the Moon, and what we see from Earth's surface is a lunar eclipse. The bright face of the full Moon (a lunar eclipse always takes place at full Moon) is bitten into by Earth's shadow. If the Sun passes behind the Earth well away from the edge of Earth's disc, the entire face of the full Moon is hidden.

How does this appear from the Moon's surface?

Well, the Sun approaches from the east, and as it approaches the Earth-crescent thins. The Earth-crescent is four times as long as the Moon-crescent we see from Earth and, moreover, is a distinct blue in color. (The light you would see shining from such a crescent would be mostly blue light scattered by Earth's atmosphere as sunlight hits it at a shallow angle, so you would be literally seeing a bit of Earth's blue sky against the Moon's black one.)

Finally, the Sun's bright disc would seem to make contact with the Earth and by then the crescent would have thinned to invisibility. The Earth would be a black and invisible circle in the sky, but its presence would be unmistakable, for it would bite into the Sun's glowing disc, much as the Moon's appears to do as seen from the Earth's surface during a solar eclipse.

In other words, what is a lunar eclipse seen from Earth's surface is a solar eclipse seen from the Moon's surface. But the Moon's version of a solar eclipse is different from ours in two ways. Its is slow-motion—it takes as much as one full hour for the Sun to pass entirely behind the Earth, and it can take up to nearly three hours before it begins to appear against the Earth's outer edge. (Compare a three-hour solar eclipse on the Moon with one that lasts for an absolute maximum of seven minutes as seen from the Earth.)

In one way the Moon's version of the solar eclipse loses out. The Earth's disc is so huge that it covers not only the Sun itself but much of its corona. As a result, the corona of the Sun is never as spectacular a sight during the solar eclipse on the Moon as it is during the solar eclipse on the Earth.

There is something else, though, that much more than makes up for this. The Earth has an atmosphere, the Moon hasn't. When the Sun is behind the Earth's disc, its light shines *through* the atmosphere all around the Earth. Most of that light is absorbed or scattered by the atmosphere, but the longest light waves survive. This means that the invisible black circle of the Earth's disc is surrounded by a rim of bright orange—what we are seeing is, in effect, a curve of sunset all around the Earth.

Picture, then, the sight of the solar eclipse as seen from the Moon. The black sky is covered with a powdering of stars much more thickly than here on Earth, and somewhere in that sky is a perfect circle of orange light beyond which is what can be seen of the pearly white of the Sun's outer corona.

And the surface of the Moon itself is lit for a while not by the harsh and brilliant white light of the Sun, and not by the cool and soft white reflected light of the Earth, but by the dim and orange light of another world's sunset.

Is this just imagination? Not at all. We can actually *see* that sunset light from the Earth, for during a total eclipse of the Moon, we generally don't see the Moon disappear. It remains perfectly visible, shining with a dim copper color in the distant sunset-glow.

It is the solar eclipse by Earth that is the supreme sight of the lunar skies. That is what the tourists will wait for confidently, since the moment of such eclipses can be predicted centuries ahead of time.

Some things cannot be predicted, however. It can be that at the moment of the eclipse, those sections of the atmosphere rimming the Earth are unusually full of clouds so that little light will get through. The orange circle will be dim, or incomplete, or even virtually absent, and the tourists will be disappointed. (At certain rare occasions, the fully eclipsed Moon *does* just about disappear, and we know the distant sunset circle has failed.)

Will there be "eclipse insurance" taken out by tourists travelling to the Moon, to guard against total loss of passage fare in case of this happening?

But do the glories of the lunar sky utterly pale the scenes available to Earthmen who won't live to reach the Moon—or won't care to make the trip? What is left in Earth's sky the Moon cannot match?

A shooting star, perhaps? Many meteors hit the Moon, but they must pass through atmosphere if they are to glow.

The beauties of our sunrise and sunset depend upon the presence of an atmosphere, and the same phenomena on the Moon are dull and colorless in comparison.

Then there is the ever-changing cloud patterns in the sky; the mist, the fog, the rain, the snow. None of this ever happens on the Moon.

There is even the sight of the calm, deep, unbroken blue of the sky of a peaceful summer day, when a person can find himself in open air stretching for endless miles in all directions and with no need for any protective garment or any curving dome to protect him against the environment.

We have all about us the infinite disregarded wonder of the Earth which, if we had the sense we apparently lack, we would labor to preserve for the priceless and irreplaceable heritage it is.

60

THE SOLAR SYSTEM
FOR HUMANITY

There were two remarkable revolutions in our study of the Solar System, each one of which flooded us with information we never dreamed we would gain.

The first came in the early 1600s, when Galileo devised his telescope, turned it on the heavens, and brought the planets nearer.

The second came in the 1950s, when we learned to send radio waves and rockets out to the planets.

The revolutions are remarkably different in one way. Galileo's telescope was homemade and cost little. The devices used to make the second revolution possible are, on the other hand, extremely intricate, extremely expensive—and the expense comes, for the most part, out of the public purse.

Of what use is the new knowledge of our planets?

Astronomers may scorn such a question. Curiosity about the universe, the desire to know the workings and fundamentals of every aspect of this huge cosmos, is the noblest aspiration of the human mind, and the knowledge gained is its own reward.

So it is, but while it is the astronomers who aspire and the astronomers whose curiosity is sated, it is the public who pays. Surely the public has the right at least to ask the question: Of what use is it?

Knowledge always has its use. However arcane the new items uncovered, there will always be a use eventually. However little scientists may be motivated by the need for a use, that use will show up. For instance . . .

1. THE WEATHER

There is no need to belabor the point that humanity is at the mercy of hot spells, cold spells, droughts, floods, monsoons, hurricanes, tornados, and every other trick our atmosphere can play. It would be good to know how to understand, and therefore predict, all the vagaries of the weather, pleasant and unpleasant. With enough understanding, a certain amount of control might become possible, too.

Prediction remains inexact, however, and understanding very limited. Even the use of weather satellites, though flooding us with hitherto undreamed-of quantities of information, and though helpful in many respects, has not rendered atmospheric movements less complicated or significantly easier to understand.

The trouble is that our planet offers a remarkably complex system. There is both land and water in irregular distribution, unevenly heated by the Sun, and the whole in fairly rapid rotation. If only we could study a simpler system with an air circulation we could more readily understand, we might extrapolate patterns from that one which might explain the situation as it is on Earth.

But there *are* simpler systems—no less than three of them nearby—and we are beginning to study them.

Consider the planet Mars. It, too, rotates on its axis in just about twenty-four hours, but because it is smaller than Earth its surface moves at only a little over half Earth's speed. It is half again as far from the Sun as Earth is, so that the heating of its surface is not as greatly uneven as that on Earth.

Finally, there are no open bodies of liquid on Mars. It is completely land, with only the minor complications of polar ice caps, slowly shrinking and growing with the changing seasons.

Next, consider Venus. Like Mars, Venus had no liquids on its surface. Nor does it have ice caps. It is a bare ball with a rocky surface. Furthermore, Venus is very evenly heated. Its entire surface is at roughly the same, very high temperature, day and night. And on top of that the rotation rate of Venus is very slow. It takes eight months for Venus to make *one* turn about its axis. All these things subtract from the complexity of the situation.

Finally, consider Jupiter. It has no solid surface at all. It is all ocean under its atmosphere (a peculiar ocean of hot liquid hydrogen, but an ocean). Its atmosphere is whipped about the planet by a speed of rotation much higher than Earth's, and because Jupiter is quite far from the Sun, Jupiter is much more evenly heated than Earth is.

To summarize:

Earth: solid and liquid surface, uneven heating, fast rotation.

Mars: solid surface, uneven heating, fast rotation.

Venus: solid surface, even heating, slow rotation.

Jupiter: liquid surface, even heating, superfast rotation.

If gravity plays a role, there is a spread there, too. Venus has a gravitational pull on its surface very much like Earth's; Mars's pull is distinctly less than Earth's; Jupiter's pull distinctly more.

It is only in the last twenty years that we have begun to learn the details of these other atmospheres. But we must know much more, because the knowledge we gain should add enormously to the understanding of our own. It may sound odd, but to really understand our weather, we may have to study the other planets; and if the other planets really help us then even a huge expenditure will be justified.

2. THE SUN

Even more than the weather, the Sun is of prime importance to mankind. A small solar hiccup, a slight increase or decrease in radiation intensity, for example—something totally unimportant on the Sun's own scale—might be catastrophic for Earth and might conceivably wipe out life altogether.

Can it possibly happen? Surely the Sun is well-behaved and steady, year after year, age after age.

Perhaps not. Our studies in recent years show disturbing indications that the Sun is not quite as sober and reliable a furnace as we might have thought.

The sunspots come and go, and the Sun is more active and a trifle warmer at sunspot maximum than at sunspot minimum—but the cycle is not quite regular and we don't know what influences it. Within the last few years, in fact, we have gathered evidence to the effect that there are periods when the sunspot cycle ceases altogether, when virtually no sunspots appear. The last of these interruptions lasted from 1645 to 1715 and coincided with what is called "the little ice age," when temperatures dropped, growing seasons shortened, and crops often failed.

In recent years, we have measured the rate at which tiny particles called "neutrinos" emerge from the Sun and we find that they reach us at only one-sixth to one-third the rate astronomers had confidently predicted they would. Nor can astronomers figure out what can be going on at the center of the Sun to account for the "mystery of the missing neutrinos."

There is even a recent suggestion that the dinosaurs became extinct (along with seventy-five percent of all the species of animals then existing) because of a minor explosion in the Sun. Material from the explosion finally reached Earth and left its mark in an increased level of the rare metal iridium, which is more common in the Sun than in the Earth's crust.

Is there any way we can learn more about the nature of the Sun in order to be able to understand its mysteries better and to predict its instabilities more accurately?

Well, we need to know the exact details of what changes are going on at the center of the Sun. We have some idea now of what its chemical composition is, but we have no sure notion of what its chemical composition was at the time it was first formed nearly five billion years ago.

In all that time, nuclear fusion has been going on in a complicated fashion, changing the chemical structure of the Sun. If we knew not only the present chemical structure, but the original chemical structure as well, we could work out more exactly what changes must have taken place and, from that, deduce the intimate details of nuclear fusion at the solar center better than we now can. This in turn might lead to a better understanding of the Sun and its irregularities—and almost no solar incident will be as disastrous to us as it might be, if we can predict it and prepare ourselves for it.

The planets were formed out of the same cloud of dust and gas that the Sun was, but in most cases the gravitational pull of the planets was not great enough to hold all the substances of the cloud completely. Earth, for instance, retained none of the helium and almost none of the hydrogen of the original cloud, and those two elements made up perhaps ninety-nine percent of the cloud. Earth's composition, therefore, gives us no hint of the composition of the original cloud.

But what about Jupiter? It is the largest of the planets, and has enough of a gravitational field to hold on efficiently to everything in the original cloud—even hydrogen and helium. In fact, all our indications right now tell us that Jupiter is a huge spinning ball of liquid hydrogen, with some admixture of helium and lesser admixtures of other elements.

Furthermore, Jupiter is not large enough to have developed nuclear fusion at the center, so no significant changes have taken place in its composition with time. It is quite likely that Jupiter is the best example we have of a true sample of the original matter of the Solar System, unchanged in all the nearly five billion years it has existed.

Thanks to our rocket probes we know enormously more about Jupiter than we knew ten years ago, but if we continue, and learn the intimate details of its chemical structure (surface *and* interior), we might well be able to understand the Sun far more than we do now, which could help us—perhaps—save our civilization from disaster in times to come.

3. LIFE

What else besides weather and the Sun concern us? What about our own bodies? What about the nature of life? of aging and death? of sickness and health?

We study our bodies with infinite care and have developed numerous techniques in recent decades for probing deeply into the smallest recesses of the cell.

There are two problems, however, which make life difficult for biologists. First, we have only one sample of life. Every one of the couple of million species of life on Earth is built on the same basic biochemical plan. If there were other basic plans to study, we might, by comparing them, learn more about each (our own included) than we could possibly learn without such a comparison.

Second, life is a very complex phenomenon. The simplest cell is a structure of enormous complications—billions of different kinds of molecules in intricate interrelationships. If only we could find some sort of pre-life; if only we could study the scaffolding or the preliminary sketches, so to speak, we might then be able to make better sense of the finished picture.

That is one of the reasons why scientists have been excited about the possibility of finding life on other planets. Those would be examples of life built up without reference to life on Earth, and how from Earth's life they would be bound to differ in at least some basic ways. These differences could be extraordinarily illuminating.

It would not be necessary to find intelligent life, or even life that was visible to the naked eye. Bacterial cells are complicated enough. If we could find the equivalent of bacteria on another world, it would do us as much good as if we had found the equivalent of an elephant.

Bacteria are extraordinarily adaptable and tenacious of life. On Earth, they are found under the widest range of environmental conditions, so it seemed not beyond the bounds of possibility that bacteria might be found in secluded spots on even so apparently lifeless a body as the airless, waterless, hot-and-cold Moon.

Unfortunately the Moon yielded us nothing. No signs of present or past life, no matter how simple, were found. The Moon is utterly dead and always has been. That was a deep disappointment.

Mars seemed a better bet—the best bet in the Solar System, in fact. Human beings have neither examined it directly nor brought back samples of its surface, but we have sent probes. They have landed on its surface and tested it.

The results are equivocal. Some interesting effects have been detected that we might expect, were there simple life in the soil. On the other hand, no organic matter worth mentioning was detected—no carbon compounds—and we don't see how there can be life without that. We need a closer look, either with more advanced machinery or with human beings actually present. However, the odds would now seem to be that there is no life in the Solar System outside of Earth.

But what about the second hope—not life, but pre-life? On Earth, before even the simplest life form evolved, there must have been a period of "chemical evolution," when molecules grew more and more complicated on their way to life. Not only is the route to life interesting in itself, but such molecules and simple molecular systems might represent the scaffolding or the preliminary sketches that we need to help us make better sense out of the living cell.

No such molecules have been found on the Moon. Nor do they exist on Mars if we can trust our long distance analyses. Are we stymied?

Wait. There is some planetary material that actually reaches us, that passes through the Earth's atmosphere and lands on the surface that can be investigated by human beings. They are the meteorites.

Most meteorites are composed of metal or rock and have nothing of interest for the biologist. A few, however, a very few, are dark "carbonaceous chondrites" which contain the light elements such as carbon and water, and even simple organic compounds that represent the very first and most primitive stepping stones to life.

There are, however, so few carbonaceous chondrites for us to study. They are brittle and fragile objects that are not often likely to survive the flaming and traumatic journey through the atmosphere. What if we could find such carbonaceous chondrites in space, where they have been undisturbed for over four billion years, and see what has been built up in them during those years in which they have been drenched in the energy of sunlight?

Easy to say, but where are they?

It has come to seem, in recent years, that at least half the asteroids, perhaps even as many as four-fifths of them, are extraordinarily dark in color. They have the color characteristics of a carbonaceous chondrite and are very likely just that, or at least have surfaces of that nature.

The only trouble is that the asteroids are farther off than Mars is, and they are so small that landing test-objects on their surfaces would be tricky indeed.

Not all asteroids are very distant, of course. There are some (not many) that approach the Earth quite closely, but they have been subjected to such heat due to their proximity to the Sun that their organic compounds have long since baked out.

Yet there are two asteroids that are close but not too close. Mars captured two asteroids eons ago, Phobos and Deimos, and they now circle it as satellites. Whereas Mars is light and pinkish in color, its satellites are dark indeed.

We can reach them. Since we can send our Viking probes to Mars, we can send them to the satellites with equal ease. On those satellites, it is likely we will finally find the organic compounds we seek, and find them in quantity, for they have built up slowly over billions of years, with no external disturbance whatsoever.

What can they teach us? How can we tell? If we knew what they could teach us, we wouldn't have to seek the satellites out. If the compounds are there, however, they will surely teach us *something,* and whatever that something is, it will tell us much about life and about ourselves.

This by no means exhausts the list of what the new Solar System means to us and can mean to us in the future. But if the three possibilities I mention here were the only ones, we would still have to decide that our investment in the exploration of our sister planets is a wise one.

Afterword: The previous essay was written in 1979. I mention the increase in the concentration of iridium in the Earth's crust at the end of the Cretaceous Period, something that had just been discovered, and say that it might have brought about the end of the dinosaurs. However, I blame it on the possibility of a minor explosion on the Sun, which was, indeed, one of the early speculations. The more firmly founded possibility of an asteroidal and cometary impact came later, after this essay was written.

61

THE CLINICAL LAB
OF THE FUTURE

As far as we know, only the human species can anticipate the future beyond the next meal or the safety of the next refuge. And it is only during the last two centuries that scientific advance has been fast enough to allow us to anticipate startling changes of that sort within the individual lifetime.

What, then, can we expect in the next thirty years?

Maybe nothing! There is always the chance of that. If humanity decides to play with nuclear bombs and to ignore the various vital problems that now face us, we may not advance at all. We may collapse.

If we are cautious and forethoughtful, however, we might go on safely and constructively. In that case, the clinical lab may take advantage of the two enormous revolutions that we face—that of space and that of the computer.

Suppose, for instance, that we have a laboratory in space. It is not impossible that this might happen within thirty years, if the major nations of the world invest, one-tenth as enthusiastically, one-tenth the money they put into their competing war-machines.

A lab in space will have available to it conditions totally different from those on Earth. In space there are high temperatures and low temperatures for the asking, depending on whether material is exposed to the Sun or shaded from it. Along with high temperature is hard radiation. While in free fall, there is zero gravity. And there is hard vacuum, as much of it as is needed.

These conditions, when they can be duplicated on Earth at all, can be duplicated only over small volumes, for limited times, and at great cost in energy.

To what use can these special conditions be put?

As examples . . . . With all the hard vacuum and low temperature you need, molecular distillation can be carried out with unprecedented ease, biological substances can be purified to additional orders of magnitude, and cryogenic work can become easier and cheaper.

But why make lists? The most exciting possibilities can't be listed, for the one thing that is absolutely certain is that there will be surprises. Clinical chemistry will take unpredicted turns that will seem inevitable only after they have taken place and been understood.

Consider, too, the fact that in space, clinical chemists are far from Earth.

There is laboratory work that can be carried out on Earth only with risk—sometimes seemingly unacceptable risk. What if pathogenic bacteria escape from the laboratory? What if radioactive isotopes are unwittingly carried off? How does one handle radioactive wastes, for that matter?

A laboratory that is separated from Earth's teeming surface by a thousand miles or more of clean vacuum is, to a large extent, rid of those fears. Accidents may happen to those in the lab, but they are volunteers who know what they face and what to do. Dangerous tests and dangerous experiments can be undertaken, and through them clinical knowledge will be advanced.

In space or on Earth, one of the most important tasks facing the clinical chemist of the future is to learn more about the complexity of the genetic apparatus.

Chromosomes consist of long chains of nucleic acids that are in turn built up of long chains of nucleotides. Those nucleotide chains govern the characteristic patterns of the enzymes that determine individual cell chemistries.

We already have the techniques necessary to work out the exact order of nucleotides in particular molecules of nucleic acid, so the art of mapping the genes of individual human beings is within our grasp.

We will be entering the stage when it is not just the fingerprints of each individual that we will be able to record, but the "geneprints." It would not be a matter of identification only, either, though nothing could be more refined or less likely to yield a false positive (except among those of identical multiple births). It would be a matter of watching for congenital disorders at birth and of searching for ways of correcting them by direct gene-adjustment.

Indeed, it will not be a matter of individual genes alone, but of combinations. As we collect more and more geneprints and correlate them with the properties and characteristics of the individuals possessing them, we will learn more and more how genes affect, cooperate with, and inhibit their neighbors. We will have "holistic molecular biology."

This new field will be far too complicated for unaided human analysis—and that's where the computer revolution will come in. It will certainly take cleverly programmed computers to make sense out of permutations and combinations unimaginably greater in number and complexity than those in a mere game of chess. Perhaps they will be able to begin to pinpont potential talent, or mania, at birth—its kind and degree.

Further still . . . the number of gene variations and combinations that have actually occurred in all human beings who have ever lived (even in all organisms that have ever lived) is virtually zero compared to the number that can potentially exist.

If we learn enough, it may be possible (with the help of computer analysis) to weigh carefully the potentialities inherent in gene combinations that may have never occurred in human beings, and gene varieties that have never actually existed. We may build up, in theory, human beings that have never lived, who possess personalities, creativities, talents, and genius of types we have never encountered.

There may be space labs devoted to the development of carefully designed fertilized eggs, and to the determination of what actually develops. In short, we would have the opportunity to direct evolution with knowledge of what it is we're doing and why—enough knowledge, at least, to allow us to take reasonable chances and expect reasonable results.

Playing God? Of course. But every time a clinical test helps save a life, we're playing God. Do *we* have a right to decide whether a human being should live or die?

Would these new abilities be put to "evil use"?

We can hope not. For again, the same growing ability to handle complexities with the aid of computers may make it possible for us to plumb the devious versatilities of the neuronic pathways of the human brain.

If we could learn the detailed workings of the brain and pinpoint the biochemical and physiological source of reason and emotion, we might learn to detect, diagnose, prevent, and cure mental abnormalities, aberration, and unease.

And with these undesirable components removed from our species, it may be that the new and extremely powerful techniques developed to modify and adjust human beings will *not* be put to "evil use."

In such correction, may we not be correcting only what we choose to define as "abnormalities"? And may we not actually reduce humanity to a dull uniformity and conformity that will squeeze out all that is worthwhile?

The danger exists, but it might be avoided. It is possible, after all, that we will move against apparent mental disorder reluctantly and minimally, and that we can then save variety and creativity, while wiping out disease.

Dryden said, "Great wits are sure to madness near allied, / And thin partitions do their bounds divide."

Yet the partitions, though thin, are not invisible, and it should be possible to distinguish the eccentricities of genius from those of madness—at least often enough to protect humanity without costing it too much.

And will all this happen in the next thirty years? If we are careful, we should at any rate see the start of it happen, and reach the point where the rest will be easily anticipated.

———— 62 ————

THE HOSPITAL OF THE FUTURE

About a century and a third ago, hospitals in scientifically advanced portions of the world were beginning to make use of anesthesia and antiseptic procedures. It was only then that they began to fulfill their modern function of helping patients recover from illness.

Ever since then, hospitals have been becoming steadily more expert in fulfilling this function—and steadily more expensive. That the two, expertise and expense, don't just go together in the dictionary would seem inevitable. Nevertheless, the problem of medical costs is becoming a national concern and, to some, a national scandal.

In assessing the future of the hospital, then, we seem to face a total impasse. If medicine is to continue to advance, it will have to become still more expensive; yet in response to rising public insistence, it will have to become less expensive. It would seem inevitable, then, that the highest pressures, and therefore the surest changes, will take place in directions that will tend to resolve this apparent irreconcilability.

The clearest way of doing so is to use high-technology devices to decrease the length of the hospital stay, so that the increasing expense of the former will be overbalanced by the decreasing expense of the latter. The goal will be, in fact, no hospital stay at all. Any person who becomes ill enough to require hospitalization will already represent, to some extent, a medical failure.

Ideally, then, the medical profession must concentrate more and more on early diagnosis, even on the detection of potential disease, and take those measures best calculated to prevent the conversion of potentiality to actuality. A hospital, then, will become primarily a place where people go for routine diagnostic checking. There the full subtlety of high-technology will go into disease prevention or, at worst, treatment in the early stages when the disease can be handled most successfully, most quickly, and with the least expense.

The application of advanced technology to diagnosis came in the 1890s with the discovery of x-rays. For the first time it became possible to study the interior of the body without using a knife.

X-rays, however, don't give much detail for the soft tissues; they aren't three-dimensionally informative (until the invention of the extremely expensive CAT-scan device); and, worst of all, they are so energetic that they can induce chemical changes in the body that have a measurable chance of resulting in cancer.

We are now entering a period, however, in which nuclear magnetic resonance (NMR) can be used, and this will represent an improvement in all the ways x-rays fall short. NMR detects the hydrogen atom most easily, and because hydrogen is a component of the water and fatty molecules which are ubiquitous in the human body, soft organs can be seen in detail. Since the manner in which hydrogen atoms respond varies with even small changes in the molecules of which they are part, changes due to disease in even its earliest stages may be spotted. NMR can be so tuned as to produce pictures at any level below the surface, so that three-dimensionality is easily attained. Best of all, there is as yet no evidence (either theoretical or experimental) to show that its use has any deleterious effect on the body at all.

NMR used in this fashion gives a primarily anatomical view of the body. For a physiological view, we might turn to the blood.

To obtain a sample of blood, it is necessary to penetrate the body, but only by what is essentially the prick of a needle. Blood analysis is already commonplace and routine. Glucose, cholesterol, triglycerides, hormone levels, and so on can be worked out without trouble.

Blood, however, contains many compounds, particularly among the protein group. In all likelihood, nothing happens in the body that is not reflected in one way or another in the blood, which after all is the pathway by which the material connection among the tens of trillions of cells (whose well-coordinated activity is essential to bodily health) is maintained.

As analytical methods are refined and made more precise, and as new and more subtle techniques (including NMR) are utilized, it will become possible to draw a virtually complete profile of the blood from a single drop, giving us the chemical structure and quantity of every different molecular species it contains.

It is very likely that no two human beings (leaving out of account identical multiple births which arise out of a single fertilized ovum) will have identical blood profiles, any more than they would have identical fingerprints or gene combinations.

What's more, the blood profile will not remain unchanged with time but will vary in one of two ways. There will be a cyclic variation with time, representing circadian (and other) rhythms, something that would be very useful to physicians, since the efficacy of medication (as one out of many examples) is likely to vary with the rhythm. Secondly, there will be noncyclic changes that will reflect some abnormality in the chemical functioning of the body.

Just as physicists can work out the intricate interactions among subatomic particles from curved lines in particle detectors of one sort or another, so will physicians be able to work out the physiological and biochemical interactions in the human body from the blood profiles. The noncyclic changes in the profile will yield information as to the onset of abnormal interactions. In combination with NMR studies of internal anatomy, a disease will be diagnosed with a remarkable degree of certainty at an extremely early stage.

Nor would this have to depend entirely upon a physician's memory of what

the NMR picture of an organ, or the blood profile, was the day before, or the year before. There is no question but that the hospital of the future will be totally computerized.

The anatomical/physiological/biochemical state of each individual will be recorded at every investigation. If we assume that each person with a due regard for his own health will submit to such diagnostic procedures each year (or more frequently, if something suspicious turns up), then the computer, comparing the latest NMR results and blood profiles with earlier ones, will detect noncyclic changes. Undoubtedly it will be the changes rather than any static picture or profile that will be particularly significant. Furthermore, if profiles of different individuals are compared, ranges will be determined which will again be more significant than findings in single individuals.

In short, computers, with vast quantities of data that can be compared and analyzed in a brief period of time, will be able to turn up a diagnosis earlier and with greater certainty than a human physician will be able to do.

Nor need we stop at the level of organs and blood. Governing the total development of the individual are its genes and the nucleic acid molecules that make it up. It was only in 1953 that we learned the true structure of these fundamental molecules, and we have been filling in the details—both structural and functional—ever since.

The time will come when we will be able to map all the genes in the chromosomes of a human being, and analyze their structures as well. If we do such genic analyses of enough human beings and correlate them with the visible characteristics of each (using computers to help), we will learn which genes are sufficiently abnormal to make it likely that their owner wil have to face the chance of some particular health problem eventually. Forewarned is forearmed.

The advance of medicine and its increasing success at being able to manipulate the human organism in one way or another has brought with it increasingly controversial, and even intractable, ethical problems. We already have to face the problem of when to allow a patient to die, or when an experimental procedure of the heroic type is justifiable.

With our ability to map genes, and perhaps modify or replace them, we will face the problem of how far we can play games with the human species generally. Ought parents to be allowed to choose the sex or other characteristics of their children? Can we decide which fetuses to abort on the basis of genetic analysis? Can we synthesize new genes altogether and test them on human beings to see if we can greatly improve lifespans or intelligence?

It would seem quite possible, then, that the hospitals of the future will indeed solve the problem of skyrocketing expense by sharpening diagnostic abilities—only to find themselves facing an even higher and more impenetrable mountain range of ethical problems.

Despite the old adage that a bridge need be crossed only when it is reached, it may well pay us to do some thinking about such matters now, before the march of progress presents us with them and catches us unprepared.

MEDICINE FROM SPACE

Some day it may be possible to fill a prescription at the pharmacist and receive in return something bearing the label: "Manufactured in Space."

You will know then that the product you have bought is purer than it would have been if Earth-manufactured, or cheaper, or both. In fact, it could well be that the product could not be placed in your hands at all if the firm producing it were confined to the surface of the Earth for its manufacturing processes.

What has space got that Earth has not? For one thing, "microgravity."

A satellite in orbit about the Earth is in "free fall." It is responding to Earth's gravitational pull completely; so is everything upon it. There is therefore no sensation of weight, a sensation that arises only when one is prevented from responding to gravitational pull.

On Earth, we are prevented from responding to the pull by its solid surface. If we were in an elevator with the cable broken and all safety devices off, the elevator and we would fall freely at the same rate, and we would feel no pressure against its floor. We would be essentially weightless—for a few seconds.

In an orbiting satellite, everything is in a perpetually falling elevator, so to speak, and feels no weight except what is produced by the gravitational pull of the satellite itself, and that is so tiny as to be barely measurable even with the most refined instruments. In an orbiting satellite, then, we experience microgravity, a pull so small we can fairly call it "zero gravity."

Where does this get us? Let us turn to medicinals. The most important of all are manufactured by living tissues, including our own. Our body, for instance, constructs large and delicate molecules (mostly of the variety we call "protein") specifically designed to fulfill certain functions. They do so in ways that we can't always follow in detail yet, but about the final result there can be no doubt. They control the body's metabolism; they encourage or retard growth; they insure the proper workings of the body's immune mechanism, fight disease and foreign protein, coordinate the workings of various organs, bring about the proper blood-clotting and wound repair, stimulate the production of red blood cells or the elimination of wastes, and so on.

There are probably thousands of different molecules, varying from one another in some degree, and the slightest variation can be vital. For instance, the normal hemoglobin molecule, which absorbs oxygen from the lungs and carries it to cells in the rest of the body, is made up of several hundred amino acids arranged (like pearls on interlocking threads) in a specific order. Change one amino acid very slightly and you have the abnormal hemoglobin that gives rise to sickle cell anemia. There may be a similar tiny change that can convert a normal gene to one that gives rise to some variety of cancer.

In living tissue, these thousands of substances exist together, but the cells, and the structures within the cells, are capable of selecting that one substance they require at a particular moment, and to do so, unfailingly, even from among other substances very similar. Molecules within cells may fit each other three-dimensionally, and if one molecule isn't quite right, the fit doesn't take place. The result is similar to that in which a lock will accept and be turned by a specific key but not by a key which is very similar but not identical—except, of course, that the body's molecular mechanisms are enormously more delicate than man-made locks. The result is that only trace amounts of some substances are needed. The cells require only a relatively few molecules and can pick those few out of the vast melange, molecule by molecule.

Well, then, scientists may need particular substances for use by patients who, for some reason, cannot form any of their own, or can form them only in insufficient quantities, or who need abnormally large quantities—and the only source for these substances may be the incredibly complicated mixture of chemicals that makes up living tissue. The complex molecules very likely cannot be formed in the laboratory either quickly enough or cheaply enough, and neither can they be picked out of the mixture, unerringly, as cells themselves manage, a molecule at a time.

There are ways, of course, methods that are much less quick and elegant than those the body uses, but that manage to do the job.

For instance, there is the technique called "electrophoresis."

Complex molecules, particularly those of proteins, have surfaces covered with atomic groupings that, when in solution or suspension in water under some particular set of conditions, carry either a positive electric charge or a negative one.

The whole molecule has a net charge that is either positive or negative. Suppose a mixture of many such molecules, suspended or dissolved in water, is made part of an electric circuit. Those molecules with a net positive charge move in one direction, while those with a net negative charge move in the other direction. The speed of motion varies with the size of the charge. Even if two different molecules both have the same net charge of the same kind, there is a different pattern of tiny positive and negative charges over their surfaces, and this results in a slight difference in the speed of movement of the two.

This means that if you allow the solution to remain in the electric circuit and have a tube that is long enough and are patient enough, you will find that all the different molecules, even very similar ones, will eventually separate.

If you don't want to wait long, you can pull out a particular section of the volume of material, and have a "fraction" that contains the substance you want.

It is still part of a mixture, but one that is not as complicated as before. The molecules that are completely different have been separated out and you have a group of particularly similar ones.

You can now subject the fraction to a second bout of electrophoresis and separate it further. Eventually, you will have plucked out a reasonably pure sample of the substance you want from the mixture, however complex it may have been.

It is a tedious and imperfect method, of course, and conditions on Earth make it all the more so. The pull of gravity tends to be stronger on some of the complex protein molecules than on the water molecules that surround them, and weaker on some of the other protein molecules than on water. The result is that there is a tendency for some of the molecules to sink, some to rise, and some to clump together in blobs. The effect of gravity on objects as tiny as molecules is small and slow, but it is enough to interfere with the electrophoretic separation. Gravitational pull tends to mix what electrophoresis is trying to separate, so that the electrophoretic technique is slowed and made less effective.

Suppose, though, electrophoresis were taking place on an orbiting satellite, where gravitational effects, to all intents and purposes, do not exist. The electrophoretic separation would not be countered by undesired mixing, and would proceed more rapidly and more efficiently. There are estimates that under microgravity conditions, the quantity of substance isolated per hour can be anywhere from one hundred to four hundred times as great as on Earth.

Pharmaceutical companies, such as Johnson & Johnson, are now seriously considering setting up such space procedures and are collaborating with NASA in this respect. The initial expense will be great, of course. Putting a satellite in orbit is not cheap, and setting up an automated electrophoretic procedure is not easy. Power will have to be made available, perhaps by way of solar cells, and periodic visits to replenish supplies and collect product will be required.

Nevertheless, according to officials from McDonnell Douglas, an aerospace company cooperating with Johnson & Johnson in this project, once the pharmaceutical factory is set up in space, it might well earn back all expenses within two years, and operate at a profit thereafter.

That is the economic outlook. What about the value to people who need the medicinals, and who could not otherwise obtain it in as pure a preparation and as cheaply, if they were forced to depend on Earth-bound procedure? What about the value to scientists themselves who could use the pure substances, available in considerable quantity, to test theories of tissue function that might greatly advance our understanding of cancer, arthritis, atherosclerosis, and other degenerative diseases, to say nothing of the process of aging itself?

The absence of gravitational effects can influence other factors that may be of importance in the preparation of substances. Some molecules mix easily with water but not with oils, and some do the reverse. What's more, water and oil tend to remain separate, and if they are mixed together, they quickly separate out again, so that substances that tend to remain in one material or the other are not easily brought together so that they might interact.

The reason for this tendency to separate after mixing is that water is denser than oil and responds more strongly to a gravitational field. Water sinks and oil rises. If oil droplets are made small enough, molecular motions counter the gravitational effect, and the mixture (milk, for example) is "homogenized." Making the oil droplets small enough is a tedious process, however.

Where gravitational effects do not exist, however, as in an orbiting satellite, oil and water when stirred together form a mixture of oil and water droplets which are *not* separated out again. It is a coarser form of homogenization brought about with no trouble at all, and it may facilitate molecular changes to produce specific substances far more easily than would be possible on Earth.

Nor is the absence of gravitational pull the only unusual characteristic of space.

Here on Earth, it is possible to separate substances by "distillation." The method involves heating a complex mixture. Some materials boil more easily than others do, and their vapors come off before the others do. The vapors pass through tubes and are cooled at the other end into liquid form, and as the droplets are collected in different containers, some substances are found in one container and some in another. If the process is carried through carefully enough, and if fractions are further distilled, any substance in the mixture can be separated out.

There is a catch, though. Many molecules, especially those with large and complex molecules, do not survive heating. The molecules fall apart into small and useless fragments.

The heating is required, in part, to overcome the pressure of the atmosphere. If the atmosphere is removed to form a "vacuum," distillation can take place at lower temperatures. Producing and maintaining a vacuum is troublesome, however, and in any man-made vacuum some molecules of atmosphere remain to interfere with the distillation process.

In space, however, there is nothing but vacuum, uncounted millions of cubic miles of it—and a better vacuum than any that human beings can prepare on Earth. It may well be possible to set up distillation procedures on automated orbiting satellites which will pluck out fragile and delicate molecules at different rates, depending on their differing tendencies to vaporize in a good vacuum, and will do so without disrupting the molecules. Separations might take place that, by this technique, could not be managed at all on Earth.

There are other properties of space, too, that might be useful. The Sun is the source not merely of energy, but of particularly concentrated kinds of energy. Ultraviolet rays, x-rays, and gamma rays are all part of its radiation, to say nothing of a steady stream of charged particles, some so energetic as to amount to soft cosmic rays. These energetic forms of radiation and particles are filtered out by our atmosphere, and comparatively little reaches the surface of the Earth.

In space, however, all of it will be available, and particular forms of radiation or particles might be tuned to initiate specific chemical changes with an efficiency impossible on Earth's surface.

Again, if a satellite, or a portion of it, is shielded from solar radiation in one way or another, its temperature will drop to far below zero. Automated procedures might produce particularly fragile molecules that would be useful but could only be preserved at very low temperatures. In this direction, too, experiments might be performed, and understanding gained, that could not be managed on Earth except with extreme difficulty and great expense.

Those, then, who think of space as a vast "boondoggle" may simply be suffering from a lack of imaginative understanding. Those who feel that the money expended on the development of space capabilities might better be spent on the relief of human suffering may simply be failing to grasp the fact that the path to such relief can actually lead through space.

—— 64 ——

REVISING THE PATTERN

There is considerable overlapping in the last three essays of the book, but they represent the problems of biotechnology from slightly different viewpoints, and each one says something the other two do not.
So I include all three.

It is possible that among all the revolutions that face humanity as it attempts to expand its knowledge of the Universe, the most significant, the most hopeful, and the most dangerous, is the one that involves cellular biology.

The characteristics and abilities of human society are built upon the characteristics, abilities, and the cooperative or competitive behavior of the human beings who make it up—at least up until now, because our technology has never yet been able to eliminate the possibility of "human error."

In turn, the characteristics and abilities of individual human beings are built upon the characteristics, abilities, and the cooperative or competitive behavior of the individual cells that make them up. The characteristics and abilities of each cell are based on the characteristics, abilities, and cooperative or competitive behavior of the genes that control its chemistry. And the genes themselves are long chains of nucleotides that make up molecules of deoxyribonucleic acid or, in abbreviation, DNA.

Start at the other end, then . . .

If we fiddle with DNA molecules, it is conceivable we can learn to adjust the genes. If we learn to adjust the genes, it is conceivable we can learn to modify the behavior of cells. If we modify the behavior of cells, it is conceivable that we can alter the state of individual human beings. If we alter the state of individual human beings, it is conceivable we can build a new and better society.

Those are a lot of "conceivables" and there is, of course, danger at every step of the way.

Yet we've started. Over the last thirty years, we have learned a great deal about the detailed manner in which DNA molecules produce replicas of themselves that can be used to supply the new cells that are continually being formed—including new egg cells and sperm cells, which give rise to new individuals altogether.

We have also learned how to pry apart DNA molecules with great care at specific sites and how to then put the fragments together again in the old order, or in a new order. We can even take fragments from two different DNA molecules and put them together to form a molecule that had never before existed. It is this recombining of DNA molecules that is referred to as "recombinant-DNA" research.

This means we are slowly learning how to control the molecules that serve to control the cells. We are learning how to design new organisms—or at least organisms with new chemical abilities. Thus, by fiddling with the genes in certain bacteria, we can produce a bacterial strain that will have the ability to design protein molecules that are absolutely identical to those of human insulin—and we have actually done so.

Until now, we have had to obtain our insulin—essential for diabetics—from cattle and swine. Such animal insulin works, but because it differs in minor details from human insulin, it is possible for the human body to develop allergies to it. What's more, the supply is inelastic since it depends entirely on the number of animals slaughtered, with each having but a fixed supply of insulin.

The newly designed bacteria, on the other hand, produce the real thing and can produce it at any rate necessary, depending on how many cultures we establish and how tirelessly we can adjust them to do their work.

More feats of this nature are to be expected in the future. To be sure, there are those who fear that, quite unintentionally, strains of bacteria may be developed with fearsome pathogenic properties—disease germs that the human body has not encountered and cannot fight off—and that they will somehow escape from the laboratory and lay humanity low with a super-epidemic.

The chances of this are extremely low, and it would be sad to give up the certain benefits of recombinant-DNA research for fear of the trace-dangers of catastrophe. Far better to search for ways of reducing the danger to a still smaller level—as, for instance, by setting up laboratories in orbit about the Earth where the insulation of thousands of kilometers of separating vacuum can further protect the teeming population of the Earth.

The real benefits of recombinant-DNA research, however, have not yet even been scratched. To adjust microorganisms to produce this chemical or that, or to consume this chemical or that, is comparatively simple, like breeding cattle to produce more milk or chickens to lay more eggs.

Recombinant-DNA research can be used to do far more than that; it can study the deepest facets of the cellular machinery.

Individual DNA molecules in the cell govern the production of specific proteins called enzymes, each of which catalyzes, or speeds, some specific chemical reaction.

The chemical reactions do not, however, exist in isolation. Each influences others, and all the thousands of chemical reactions in the cell form a kind of network that is intimately interconnected, so that you cannot alter one without affecting all the others to one extent or another. (That is why all chemical treatments of any bodily disorder invariably have "side-effects.")

Further than that, enzymes, in catalyzing a specific reaction, do not work in isolation, but are themselves stimulated or inhibited by the manner in which other enzymes bring about their catalytic effects.

And still further, even the DNA molecules do not work in isolation. Each DNA molecule is influenced by its neighbors. In every type of cell, some DNA molecules (different ones for each type of cell) are totally inhibited, even though the basic ingredients of DNA molecules in all cells are identical.

Therefore, in order to understand the workings of cells thoroughly, it is not enough to consider just individual reactions, enzymes, or DNA molecules, as we have tended to do in the past out of sheer lack of ability to do anything else— but we must consider the entire overall "gestalt" of the cell.

Recombinant-DNA techniques may offer us a chance to do that, since they may make it possible to introduce minor changes in specific DNA molecules in intact cells and observe the changes in cellular characteristics that result. Changes here, changes there, combinations of changes, each one offering information: until, out of all the information, we begin to build a sense of the cellular lacework and to understand the pattern—not the individual strands that make up the separate parts of the pattern, but the *whole*.

Will this not mean that we will rapidly outpace the ability of our mind to interpret the information we get? Will not the complexity of what we learn be too much for us?

After all, there are thousands of different DNA molecules in the cell, producing thousands of enzymes, catalyzing thousands of chemical reactions. Each molecule can be changed slightly. The order of the nucleotides out of which each is built can be changed slightly or radically; and the individual nucleotide can be slightly altered even while it retains its place in the molecule. The number of possible changes that can be made in this fashion cannot be called astronomical, since there are no numbers that one meets in astronomy that are large enough. We would have to say "hyper-astronomical."

It would be necessary to simplify the problem, of course, and, in addition, to increase our own ability to handle it.

In simplifying, we would have to find key changes in our DNA manipulation. After all, not all changes produce really interesting results. If one were dealing with a huge factory, knocking out the change-giving facility on the coke machine would alter events in the factory far less than distorting one of the intercom devices would. By searching for key changes and concentrating on those, the complexity would be reduced from the hyper-astronomical to the merely astronomical.

To increase our ability to handle the problem, there are computers. The human brain may not be able to handle all the variables or perform all the operations quickly enough, but a computer might, and it is its analyses we would depend upon.

We could, therefore, learn how to map the DNA molecules of a cell thoroughly and, having produced the map, learn how to understand it thoroughly, and how to consider the potentialities of the cell under the control of those particular DNA molecules. If we then graduate from microorganisms with their single cells to human beings with their trillions of cells, we would find ourselves with a new order of fuzziness (since the cells all influence each other) but with a far greater level of importance to our work.

The time will come, perhaps, when each individual will have his gene-print on record. In fact, it might be that every infant, on birth, will be routinely gene-printed in order to get a notion as to its potentialities.

This may be viewed as "playing God," as putting each person into a slot from which he will not be allowed to emerge, as a way of establishing a new and more vicious elitism.

On the other hand, it is quite apparent that no human being ever realizes his potential in the hit-and-miss treatment he gets in our present society. Tailoring education and social influence to the actual potentialities of each may allow all human beings to be far better off than without such methods. Some may be more elite than others in one fashion or another, but all will be elite compared to today's people.

Even this is not the limit. There are, to be sure, uncounted billions of different DNA molecules, differing in the total number of nucleotides, in the proportion of different types of nucleotides, and in the order and arrangements of nucleotides. There are perhaps billions of billions that have existed in all the various organisms from viruses to sequoia trees, through all the history of life on this planet over the last 3.5 billion years. Yet all of these, when compared to all the different molecules that could *conceivably* exist, shrink to virtually nothing.

Won't scientists someday, on the basis of what they learn from the DNA molecules that do exist, begin to work out tentative rules of behavior that can possibly be extended to DNA molecules that have not yet existed? If they do so, might they not learn what factors of DNA molecular structure contribute to the production of a kind of pattern in a human being that would make it easier to develop intelligence, talent of one kind or another, creativity, humor, judgment, prudence, temperance, sympathy, love?

And won't scientists someday wonder whether certain changes in the nature or pattern of extant DNA molecules might not serve to improve certain human characteristics in ways deemed desirable?

There would be a strong tendency to want to produce those DNA molecules and to insert them in human beings, except that we would scarcely dare to do so on the basis of theory alone, for what side-effects (undesirable, or even fatal) might there not be?

There would have to be experimentation, therefore, and one might imagine laboratories in orbit given over to the science of "fetology."

We might imagine endless rows of human egg cells, carefully analyzed for their gene-print, carefully modified in certain theory-directed fashion, carefully fertilized with a sperm of known gene-print that is perhaps also modified. The fetus would be allowed to develop in the laboratory in order that its properties could be carefully and continually observed. Some might be allowed to proceed to term so that actual babies would be observed; and some of these could be allowed to grow to maturity, when that would be necessary to test the theories.

Is such a science, and are such experiments, repugnant? They are to me. But then, animal experiments are repugnant to me, too, yet there is no way of doing without them, so far, if medical and biological research is to advance.

What's more, human attitudes change. Dissection of human cadavers was once forbidden because it meant the desecration of the human body, but medical knowledge could not advance under that prohibition, and uncounted human lives must have been lost out of this exaggerated respect for the human dead.

It may be that eventually people in general will recognize the importance of fetology to the survival of the human species. It is constantly being said that human knowledge has outstripped human wisdom, and that machine-control has advanced beyond self-control; and that in this disparity of development lies the dismayingly huge chance that we will destroy ourselves.

Well, perhaps we can stave off destruction until we have learned enough about the pattern of the human body to devise new patterns less likely to bring about that destruction. Perhaps we can learn to guide human evolution and to do so in the direction of better-than-human.

Cynics may say that even if we learn to do so, the worst facets of human behavior will guide that evolution toward the self-serving benefit of the few, and that the latter end will be worse than the beginning.

Maybe so, but I don't quite hate and despise humanity to such an extent I feel there is at least some chance for us to learn to better ourselves, and honestly to strive to do so. And if there is such a chance, then it seems to me we ought to try for it.

65

PUTTING BACTERIA TO WORK

Diabetes is the most common metabolic disease, and once there was no known way of successfully treating it. It was in the 1920s that the cause was found to be the lack of a pancreatic hormone we now call "insulin." Diabetics could not form insulin in their own pancreas glands but could lead a reasonably normal life if they were supplied with insulin from some other source.

Till now, the source has been the pancreas glands of domestic animals; cattle and swine, particularly. Each slaughtered animal has one pancreas, and from the pancreas, insulin can be isolated.

The trouble is that as the population increases, so does the number of diabetics, while the supply of insulin is sharply limited by the number of animals slaughtered. Furthermore, insulin from cattle and swine is not *quite* identical to human insulin. Animal insulin is close enough to work, but there is always the possibility of developing allergic reactions to those alien molecules.

Hence, the excitement over the fact that *human* insulin is now on the market. What's more, it has the potential to be produced in almost unlimited quantities. And it doesn't come from human beings—or from any animal. It can be manufactured by bacteria.

How did that come about? It began in 1973, when two Californian researchers, Stanley Cohen and Herbert Boyer, were working with a common intestinal bacterium, *E. coli*. They perfected a new technique for dealing with DNA (deoxyribonucleic acid), those molecules within cells (human, bacterial, and everything in between) that serve as blueprints for the manufacture of specific proteins.

DNA molecules exist in long chains or, sometimes, rings, and these may contain hundreds or thousands of sections (or "genes") each of which controls the production of a particular protein. Cohen and Boyer worked with two kinds of enzymes. One kind can cut a DNA molecule in a particular place. Different kinds of such "restriction enzymes" can cut it in different places. A second kind of enzyme, "DNA ligase," can bind the pieces together again—recombine them.

If two different strands of DNA molecules are split into pieces, and if the various pieces are mixed, and then the DNA ligase is used, the pieces are built up into long strands or rings again, but not necessarily in the original form. The recombined strands or rings of DNA ("recombinant-DNA") contain pieces from each of the two different strands and make up a new strand not necessarily exactly like any other strand that exists in nature. The DNA has been recombined into something that is, very likely, completely new.

Bacterial cells can sometimes live with these recombinant-DNA molecules. As the bacteria grow and multiply, the changed molecules are copied and each new bacterial cell has them. In this way, a goodly quantity of the recombinant-DNA is produced or, as it is sometimes referred to, "cloned." All this is an example of "biotechnology."

Naturally, since the recombinant-DNA is not quite like the original molecules, it produces proteins that are not quite like the originals.

This, obviously, has its dangers, and the early researchers in this field were well aware of them. Suppose a gene is produced that can in turn produce a protein that is a poisonous toxin. Suppose that the *E. coli* that has been changed into such a toxin-producer gets out of the lab, somehow, and into human intestines. Suppose it then produces a "super-plague" against which we have no natural defenses. The thought of accidentally wiping out the human race was so frightening that for a time scientists themselves led a movement for sharply restricting research in this field, and for setting up enormously strict precautions for such research as was allowed.

It turned out that the fear was an exaggerated one. The bacterial cells that contain recombinant-DNA are, in a sense, abnormal, and can only be kept alive through painstaking effort. Furthermore, the chances of producing a dangerous gene by sheer chance is incredibly tiny. The thought of the usefulness of biotechnological techniques, of the help they can give us in a hundred different directions, outweighs by far the considerations of a nearly nonexistent danger.

Nor are scientists "playing God." At least, if they are, they have been doing it throughout history.

Ever since animals and plants were first domesticated, ten thousand years ago or more, human beings, knowingly or not, have been tampering with genes.

They have preserved those animals they found most useful; cows that gave much milk, hens that laid many eggs, sheep that produced a thick fleece of superior wool, strains of wheat or barley that produced numerous plump grains on each stalk, and so on. They bred those to the exclusion of others so that the domesticated plants and animals of today are enormously different from the original wild species that prehistoric men first dealt with. These plants and animals have been changed so as to be more useful to human beings, not necessarily to themselves, and this change has been brought about by the slow and clumsy alteration of genes through evolutionary mechanisms—though herdsmen and farmers did not know that that was what they were doing.

Even work with microorganisms dates back to prehistory. Yeasts and molds were used by human beings (who knew nothing about their cellular, let alone molecular, nature) to ferment fruits and grains, thus producing wine and beer; they were used to form light and fluffy "leavened" bread and pastry; they were used to form cheeses of various kinds.

The difference is that biotechnology takes us down to the level of the molecule so that we can *design* genes that will produce what we want. We can take a particular gene and incorporate it into a bacterial chromosome, which will then produce what that particular gene naturally produces. Or we can use new tools to analyze the precise structure of a particular gene, find a similar gene in a bacterial chromosome and then modify it to change the similarity into an identity. Or we can build the gene we want from scratch and place it in the bacterial chromosome.

However we work it, we end up with a bacterium that is a cellular factory producing something that *we* want, even though the bacterium may have no particular use for it. (Are you sorry for the bacteria? What use has the hen for the quantities of infertile eggs it lays for our breakfast table?)

Thus, when a human gene that manufactures the human variety of insulin is inserted into a bacterial cell, that cell will multiply into uncounted millions, if we allow it to do so, and all of them will manufacture the insulin they have no use for themselves. And we can then harvest it, and, indeed, *are* harvesting it.

Human insulin is merely the first of the bacterially produced molecules that has become a commercial product. There are numerous other molecules that are on the verge of similar success.

One example is "interferon." This is a molecule which is used by the human body as a defense against viruses. Unfortunately, it is very difficult to isolate from human tissue (which is enormously complex, of course). If, by the proper biotechnological techniques, we can devise a strain of bacteria that produces interferon in quantity, we will undoubtedly be able to isolate it without undue trouble.

If we have quantities of pure interferon to work with, we may well find out the conditions under which it is most effective in dealing with a wide variety of virus conditions—and this means from the lowly wart to cancer itself.

Again, there is human growth hormone, the lack of which results in severely undersized human beings, and the use of which in early life might produce normality. Bovine growth hormone (produced by cattle) won't work on human beings, but careful doses seems to increase the milk production of cows.

And the same is true of a variety of other hormones, antibodies, enzymes, and so on. All of these are proteins of one kind or another; all can be obtained, by older methods, in tiny quantities after enormous trouble, so that their undoubtedly useful properties are drowned in the impracticality of the time and labor required; and all of which might be obtained in far less time, in far higher quantity, and, very likely, in far greater purity, by the use of biotechnological techniques that produce properly designed bacteria.

There are still dangers, of course. Even if we dismiss the dramatic possibilities of a new Black Death wiping us all out, there is always the temptation to misuse the powerful proteins that might suddenly become available in quantity.

One example . . . . A human growth hormone would make it possible for a child with an insufficiency of that hormone to grow normally, and that would be good—but unless his (or her) own genes were remodelled, defective genes could still be passed on to offspring, and there would be a serious question as to whether there ought to be any.

Again, there might be the temptation to use the hormone on normal human beings merely to make them a little taller, because tallness might make it possible for them to earn a better living as a model or a basketball star. The trouble is that an oversupply of a hormone is as dangerous as an undersupply, and that the side-effects can be serious indeed.

But then, it is not just these powerful proteins that could be the products of the bacterial factories. We might work our way down to much more prosaic products.

It is conceivable, for instance, that we might endow our bacteria with the kind of genes that produce muscle proteins. We might then harvest bacteria not for the purpose of obtaining this molecule or that but for obtaining a bacterial mass that tastes and feels like meat.

Why bother? Why eat bacterial meat when we can have the real thing? Ah, but will we have the real thing? It is to be hoped that the human population will level off and reach a plateau in the twenty-first century, and then decline gently. Meanwhile, though, we will have to feed people at that plateau, and bacteria will be able to produce meat protein far more quickly than cattle or swine can, and do so without extraneous material such as gristle and bone, to say nothing of starting with "food" that would be of far less value than that which animals would require.

Or consider the possible uses of biotechnological techniques in industry. It is possible that we may design bacteria that produce large quantities of methane—the key component of natural gas. Then, too, there has been speculation to the effect that the proper kinds of bacteria can act to force oil out of their adherence to rock particles and thus bring a larger percentage of the contents of an oil well to the surface.

Consider, also, that cells normally make use of small quantities of certain elements that are present in the environment in smaller concentrations still. Cells must extract and store such elements. Bacterial cells may be designed which can do so with greater speed and efficiency than untouched cells would, and that might do it in the case of elements other than those they need for their own use. In this way, bacteria can serve us as tiny efficient miners, and ores that are far too lean for us to treat economically by ordinary chemical methods will become vital sources for metals of all kinds.

Here's something else . . . . Bacteria and other microorganisms can break down molecules of all kinds. We might design microorganisms that can live on pollutants which otherwise must accumulate dangerously. (Caution! A pollutant in one place is a valuable resource in another. If we have molds that can break up discarded plastic containers, we will have to work out ways of preventing them from attacking plastic containers *before* we discard them.)

Finally, we must not underestimate the value of biotechnology in basic research. By designing and studying genes in detail, and by making use of the enzymes, hormones, and other molecules that biotechnology will provide us in quantity and in purity, we may well learn the exact details of certain processes in living cells which at present still remain obscure to us.

The example that must spring to the mind at once is cancer. Despite decades of devoted labor, scientists do not yet know exactly what it is that goes wrong in a cell and causes it to begin undisciplined growth. If we *could* find out, the cure, or even the prevention, might become obvious.

Again, there is the process of photosynthesis, which in the chloroplasts present in the cells of green plants splits the water molecules and hydrogenates carbon dioxide molecules to starch, using as an energy source nothing more than ordinary sunlight. If we knew the process in full detail, we might be able to imitate it, or, perhaps, even improve its efficiency, and we might then learn to convert sunlight into food independently.

We are at the beginning of this new science of biotechnology. Where it will lead we can only dimly foresee—and any guess may be an underestimate.

66

FIDDLING WITH GENES

Each of us grows to a certain height and then stops. Some of us may wish we had managed a couple of additional inches before stopping, but it is out of our hands.

One factor which dictates our height is "growth hormone," a protein produced by our pituitary gland. Perhaps if growth hormone were injected into a child at a young age, it might grow a little taller than otherwise. However, growth hormone is hard to get and tricky to use.

But then, the production of growth hormone is controlled by a particular gene in the chromosome of the cells, and in that sense the amount produced by a particular individual is determined from the moment of conception. Might not the gene be altered, somehow? Might not a different gene be substituted?

You can't very well experiment with human beings in this respect, but you can with animals.

For instance, the various kinds of rats and mice belong to related families, but on the whole rats are considerably larger than mice. Presumably, growth hormone is more effective in rats, or exists in greater quantity, and a difference in genes may be responsible. What, then, if the appropriate rat gene were injected into the developing mouse egg cell?

This has been tried recently and, in some cases, the egg cell so injected resulted in a baby mouse that grew extraordinarily quickly for a mouse. At an age of ten weeks, an ordinary mouse might weigh three-fourths of an ounce; the mouse with the rat genes, however, weighed one and a half ounces, or twice as much. Except for that, the giant mouse was completely mouselike.

Gene transfer is not the only way in which scientists can fiddle with genes.

There are, for instance, two genes in human beings which control the formation of hemoglobin, the protein in red blood corpuscles that absorbs oxygen in the lungs and carries it to the cells. One gene is active at the time a human being is developing in the womb. It produces "fetal hemoglobin." Once a baby is born that gene is switched off and the second gene, which produces "adult hemoglobin," goes to work.

Adult hemoglobin ordinarily does a better job than fetal hemoglobin but sometimes that second gene is defective. In such cases, a somewhat abnormal hemoglobin is formed and, as a result, a person may suffer from "sickle cell anemia" or from related diseases. There is no way of curing such a disease except by fiddling with the genes.

One way of doing so is to deal with that first gene, the one that produces fetal hemoglobin. The gene is still there and exists throughout life; it has just been put out of action immediately after birth by means of a small chemical change. Suppose it could be put back into action by reversing that change. Once again, fetal hemoglobin would be formed—and though it might not be as good as adult hemoglobin, it would be better than *imperfect* adult hemoglobin.

A particular chemical that was known to prevent the kind of chemical change that switched off the gene was tested on baboons. Once the baboons were seen to have suffered no ill-effects from the dosages used, it was tried on two patients with sickle cell anemia and one with a similar disease called beta-thalassemia. Some improvement was indeed noted, so that it may be that the first gene was indeed switched on again.

What do experiments such as these mean for human beings generally?

To begin with, they don't mean anything immediately. If we are at the start of an era of genetic engineering, where humanity can mold itself more closely to its heart's desire, it can, and probably will, take a long time to convert that start into practical, everyday medical treatment. Much remains to be done.

After all, once the Wright brothers flew their first canvas-and-pianowire airplane at Kitty Hawk in 1903, we could not suppose that the next year we would be carrying a hundred passengers across the Atlantic at supersonic speeds. And once Robert H. Goddard flew his first liquid-fuel rocket a mile into the air in 1926, we could not suppose that the next year we would be taking closeup pictures of Saturn. In both cases, a half-century of intense and ingenious invention and development was required, and—inevitably—some casualties. . . .

And genes, remember, are far more complex than planes and rockets, and far more intimately involved with each of us, so that the risks accompanying error are greater and more frightening.

As it is, the successful experiments recently conducted are only limited successes at best. Injecting foreign genes into a developing egg cell is not an easy procedure. Less than ten percent of the mouse egg cells injected with rat genes developed at all, and of those that did only a minority showed the remarkable growth effect.

Again, in switching on the gene for fetal hemoglobin, the chemical used seemed to work, but it was a toxic chemical. It might be used once or twice without damaging the patient more than the disease did, but regular use would surely kill him.

Suppose, though, that the necessary advances are made over the next half century or so. Suppose that better techniques are found for gene transfer and gene manipulation. Suppose we learn how to turn genes on and off safely, or how to treat a gene chemically (or otherwise) in such a way as to modify its workings at will, and as we choose.

What then?

We must still be careful. The product of about 3.5 billion years of evolution is not lightly to be fiddled with.

Let us consider the possibility of added growth hormone, for instance. It seems a simple thing. If tall children are wanted, then one would add to the developing egg cells genes from tall individuals, or one would treat the egg cell in such a way as to make the growth hormone gene work longer or better.

But more growth hormone is not necessarily good. There are occasionally individuals in whom growth hormone is present in abnormally high quantity. They grow to heights of eight feet and more and suffer from "gigantism," a disorder even more serious than the dwarfism that results from an undersupply.

The trick, then, would be to induce a few added inches without bringing about gigantism, and that might be a game few would care to play. And remember that this might well be true of any form of gene manipulation. It might always be a matter of trying to place a golf ball on a very small green, where overshooting the mark would be as bad as undershooting.

Then, too, genes don't work in isolation. We don't know how a change in one gene might affect another.

For instance, the production of large mice gives rise at once to the thought that such techniques might be used on cattle, sheep, chickens, horses, and so on. In that way, we might certainly expect larger animals, but is it mere size we want? Would a larger cow necessarily give more milk, and if it did, would the milk necessarily be as rich as it might ordinarily be? Might there not be a chance that with growth hormone, a sheep would produce more but poorer wool, chickens larger but less tasty eggs, and so on?

Of course, having changed one gene, we might go ahead to change another to counteract any insufficiencies or disappointments of the first change. That, in turn, might make necessary still another change, and then yet another. Might it not be that with each additional change there is added risk?

Thus, we are now able to analyze genes in some detail, and we can compare the genes of chimpanzees with those of human beings. It turned out, to the surprise (and even shock) of those doing the investigation, that the differences are unexpectedly small. It is astonishing that differences so apparently minor can result in changes as large and crucial (to ourselves) as between a chimpanzee and a human being.

Well, then, as we change more and more genes in the hope of fine-tuning a set of improvements, might we be running the risk that after a certain point we might unexpectedly bring about a horrible change for the worse?

Let us go one step farther. Suppose we find out how to adjust genes in such a way as to achieve all the improvements we want, without introducing any appreciable disadvantages, so that individual human beings are better off in every way and worse off in none. Might there not be, even so, problems on a world scale?

Suppose we produced a world of six-foot-plus individuals, all strong, healthy, and bright. They would have to eat more, too. Unless we reduced the population, there would be an unbearable strain on our food supply.

And this has happened before. Advances in medical science in the past century have succeeded in making humanity generally healthier and in doubling the life-expectancy. This has meant an unusual increase in world population, which is now triple what it was a century ago. The increase is still continuing and is bringing enormous problems in its wake.

There is an even more subtle problem involved with gene manipulation. Let us suppose that through great skill and caution we improve lives, bringing about advances in both mind and body while making sure that we also take into account the world-wide effects of these advances. There will nevertheless surely be a tendency for some genes to be unpopular and others popular for less than vital reasons. There will be fashions and fads in genes, with enormous numbers of people insisting on fashionable genes for their children.

On the whole, this might result in certain types of genes, which are viewed as undesirable or as merely unfashionable, being wiped out. Humanity, overall, might possess a smaller variety of genes altogether. We might all become a little more similar in many ways.

This, too, is dangerous, for the existence of a wide variety of genes is a definite advantage in the evolutionary game. Some genes may seem less desirable than others, but that is not necessarily an absolute. Given different conditions, a different environmental, social, or cultural milieu, a gene that seems disadvantageous may prove to have unexpected advantages.

Thus, nearsightedness may seem disadvantageous, and in youth it is. With increasing age, however, the lens hardens and is less able to accommodate close work. Under such conditions, an originally nearsighted person is better off than one with originally normal eyes is.

Consider an example involving other kinds of life. Human beings have developed particular strains of wheat that grow faster than others and produce superior flour. The result is that increasing fractions of total wheat production are confined to those strains, while other seemingly inferior strains are allowed to die out. The desirable strains, however, require a great deal of water and fertilizer—which may not always be easily available. Again, if a disease should develop that affects those particular "desirable" strains, vast quantities of grain might be wiped out, and there would be insufficient quantities of the other, supposedly inferior strains to turn to. That could bring about a world catastrophe.

If we decrease gene variety in human beings, then, we risk terrible trouble when an unlooked-for difficulty arises with which the wiped-out genes would have been better able to cope. In brief, a species without adequate gene variety has a lessened ability to evolve further and an increased liability to face extinction as conditions change.

Does all this mean that we should on no account fiddle with genes?

No, of course not. Genetic engineering offers us the hope of curing or preventing diseases such as cancer, arthritis, and atherosclerosis, which have heretofore defeated us. It offers us the hope of curing or preventing mental disorders and hormonal deficiencies. It offers us the hope of encouraging a beneficial evolution of the human species, doing deliberately and with minimal suffering what it would take "nature" millions of years to do at enormous cost. It offers us the hope of doing the same for other species and of weaving a stronger and better ecological balance of life in the world.

It is, however, vital to remember the difficulties that all this necessarily entails. At every step, we must take those difficulties into account, moving slowly and cautiously, and being always prepared to retreat at any sign that the step is a false one. The stakes are too great for anything else.

ACKNOWLEDGMENTS

"Unity" was written in 1985 for a book by United States Senator Spark Matsunaga, who used only a few paragraphs. It appears here, in full, for the first time.

"The Scientist as Unbeliever" appeared in *The Encyclopedia of Unbelief,* edited by Gordon Stein (Prometheus Books, 1985). © 1985 by Gordon Stein.

"The Choking Grip" appeared in *TV Guide,* July 18, 1981, under the title "Censorship: Its Choking Grip." © 1981 by Triangle Publications, Inc.

"Human Mutations" appeared in *Young Mutants,* edited by Isaac Asimov, Martin Greenberg, and Charles Waugh (Harper & Row, 1984). © 1984 by Nightfall, Inc., Martin Greenberg and Charles Waugh.

"The Hollow Earth" appeared in *Isaac Asimov's Science Fiction Magazine,* August 1986. © 1986 by Davis Publications, Inc.

"Poison!" appeared in *Murder on the Menu,* edited by Carol-Lynn Rössel Waugh, Martin Harry Greenberg, and Isaac Asimov (Avon, 1984). © 1984 by Nightfall, Inc., Carol-Lynn Rössel Waugh and Martin Harry Greenberg.

"Competition!" appeared in *The Science Fictional Olympics,* edited by Isaac Asimov, Martin H. Greenberg, and Charles G. Waugh (New American Library, 1984). © 1984 by Nightfall, Inc., Martin H. Greenberg and Charles G. Waugh.

"Benjamin Franklin Changes the World" appeared in *Cricket,* July 1976. © 1976 by Isaac Asimov.

"Fifty Years of Astronomy" appeared in Natural History, October 1985. © 1985 by American Museum of Natural History.

"The Myth of the Machine" appeared in *Science Fiction: Contemporary Mythology,* edited by Patricia Warrick, Martin Harry Greenberg, and Joseph Olander (Harper & Row, 1978). © 1978 by Science Fiction Research Associates and Science Fiction Writers of America.

"The Perennial Fringe" appeared in *The Skeptical Inquirer,* Spring 1986. © 1986 by The Committee for the Scientific Investigation of Claims of the Paranormal.

"The Case Against 'Star Wars' " appeared in *Newsday Magazine,* May 12, 1985. © 1985 by Newsday.

"Short Term; Long Term" appeared in *Newsday Magazine,* May 6, 1986, under

by Ziff-Davis Publishing Company.

"The Immortal Sherlock Holmes" appeared in *Newsday Magazine,* September 2, 1984, under the title "Why I Love Sherlock Holmes." © 1984 by Newsday.

"Gilbert & Sullivan" appeared in *Newsday Magazine,* November 4, 1984, under the title "An Ode to Gilbert & Sullivan." © 1984 by Newsday.

"Mensa and I" appeared in *Mensa,* edited by Victor Serebriakoff (Stein and Day, 1985). © 1985 by Victor Serebriakoff.

"Write, Write, Write" appeared in *Communicator's Journal,* May/June 1983. © 1983 by Communicator's Journal, Inc.

"Facing Up to It" appeared in *Family Health,* December 1974, under the title "Mastectomy: A Husband's Story." © 1975 by Family Health Magazine.

"Triple Bypass" appeared in *Newsday Magazine,* April 22, 1984, under the title "My Triple Bypass." © 1984 by Newsday.

"The Elevator Effect" appeared in *Science Digest,* May 1979. © 1979 by The Hearst Corporation.

"2084" appeared in *Signature,* January 1984, under the title "After 1984, What?" © 1983 by The Diners Club, Inc.

"Society in the Future" appeared in *Future,* April 1978, under the title "The Future of Society." © 1978 by Future Magazine, Inc.

"Feminism for Survival" appeared in *Science Digest,* March 1980, under the title "Wanted: Half of the Human Race to Solve Life-or-Death Crisis." © 1980 by The Hearst Corporation.

"TV and the Race with Doom" appeared in *TV Guide,* June 5, 1971, under the title "Our Race with Doom." © 1971 by Triangle Publications, Inc.

"The Next Seventy Years in the Courts" appeared in *ABA Journal,* January 1985, under the title "The Next 70 Years of Law and Lawyers." © 1985 by the American Bar Association.

"The Future of Costume" appeared in *Esquire,* September 1985, under the title "The Seers." © 1985 by Esquire Associates.

"The Immortal Word" appeared in *Writer's Digest,* May 1986, under the title "Your Future as a Writer." © 1986 by F & W Publications, Inc.

"Liberty in the Next Century" appeared in *Penthouse,* July 1986, under the title "Lady Liberty." © 1986 by Penthouse International, Ltd.

"The Villain in the Atmosphere" appeared in *Newsday Magazine,* July 6, 1986, under the title "Are we Drowning our Tomorrows?" © 1986 by Newsday.

"The New Learning" appeared in *Apple,* vol. 2, no. 1.

"Technology, You, Your Family, and the Future" appeared in *Ladies Home Journal,* September 1983, under the title "The Future Can Be Yours." © 1983 by Family Media, Inc.

"Should We Fear the Future?" appeared in *MicroDiscovery,* November 1983, under the title "Fear Not the Future." © 1983 by MicroDigest, Inc.

"Should We Fear the Computer?" appeared in *MicroDiscovery,* December 1983-January 1984, under the title "Fear Not the Future, Part II." © 1983 by MicroDigest, Inc.

"Work Changes Its Meaning" appeared in *Personnel Administrator,* December 1983, under the title "Creativity Will Dominate Our Time After the Concepts of Work and Fun Have Been Blurred by Technology." © 1984 by American Society for Personnel Administration.

"Nuclear Dreams and Nightmares" appeared in *TV Guide,* June 1, 1985, under the title "Nuclear Power." © 1985 by Triangle Publications, Inc.

"The New Tools in Space" appeared in *Popular Mechanics,* July 1986, under the title "Wings for America." © 1986 by The Hearst Corporation.

"Living on the Moon, Part I" appeared in *Food Service,* July 1969, under the title "The Food Supply of Luna City, Part I." © 1969 by Electrical Information Publications, Inc.

"Living on the Moon, Part II" appeared in *Food Service,* August 1969, under the title "The Food Supply of Luna City, Part II." © 1969 by Electrical Information Publications, Inc.

"The Skies of Luna" appeared in *Penthouse,* April 1971, under the title "A Tourist's View of the Moon." © 1971 by Penthouse International, Ltd.

"The Solar System for Humanity" appeared in *P.M.* (German language), under the title "Wie wir unser Sonnensystem für die Menschheit nutzbar machen können." This is its first appearance in English.

"The Clinical Lab of the Future" appeared in *Lab World,* December 1980, under the title "Anticipating the Laboratory of the Future." © 1980 by North American Publishing Co.

"The Hospital of the Future" © 1984 by Hospital Research Associates, Inc.

"Medicine from Space" appeared in *Health,* March 1983, under the title "Made in Space." © 1983 by Family Media, Inc.

"Revising the Pattern" appeared in *Future Life,* March 1981. © 1981 by Future Magazine, Inc.

"Putting Bacteria to Work" appeared in *Science Digest,* March 1983, under the title "The Union of Genes and Genius." © 1983 by The Hearst Corporation.

"Fiddling with Genes" appeared in *Family Weekly,* April 24, 1983, under the title "Oh, Brave New World." © 1983 by Family Weekly, Inc.